China's
Power Projection

China's
Power Projection

Air Commodore R V Phadke (Retd)

(Established : 1870)

United Service Institution of India
Air Marshal S Mukherjee Chair

Manas Publications
New Delhi - 110 002 (India)

MANAS PUBLICATIONS
(Publishers, Library Suppliers, Importers & Exporters)
4402/5-A, Ansari Road (Opp. HDFC Bank)
Darya Ganj, New Delhi-110 002 (INDIA)
Off. ✆ 23260783, 23265523, Res. ✆ 23842660
Fax: 011-23272766
E-mail: manaspublications@vsnl.com
manaspublications@gmail.com
www.manaspublications.in

2nd Impression 2011 (FP - 2005)

ISBN 81-7049-220-3

₹ 595/-

Typeset at
Manas Publications

Printed in India at
IYN Graphics, Delhi
and published by Mrs Suman Lata for
Manas Publications, 4402/5-A,
Ansari Road, (Opp. HDFC Bank),
Darya Ganj, New Delhi-110 002 (INDIA)

Contents

Acknowledgement

At the outset, I would like to place on record my deep gratitude to Professor Matin Zuberi, formerly with the JNU and Air Cmde (Retd) Jasjit Singh AVSM, VrC, VM, for initiating me into the realm of strategic studies.

Being an Indian Air Force (IAF) officer, my interest in Chinese military naturally goes back nearly four decades, but for the first twenty years of this period, it was restricted to what may be called tactical 'bean-count' calculations and purely fighting capabilities. It was when I was posted as Directing Staff (Instructor) at the Defence Services Staff College, Wellington, India, in 1984, that I began to take serious interest in China's strategic capabilities. What with frequent transfers and field assignments, this interest was often confined to sporadic reading and occasional briefings and seminars.

It was with my assignment as the AF member and senior fellow at the Institute for Defence Studies and Analyses (IDSA), New Delhi, in 1999 that I actually got the opportunity to study China in a more comprehensive manner. While at the (IDSA), its director, Air Commodore Jasjit Singh and many others including Drs. Ramachandran, Sujit Dutta, Swaran Singh, Srikant Kondapally, and Shri. M.V. Rappai, all experienced China scholars, very generously shared with me their vast experience and knowledge of the subject and gave me rare insights into the way the Chinese system worked. I am indeed grateful to them.

An opportunity for a short visiting fellowship at the Centre for International Security and Cooperation, Stanford University, California in the summer of 2001 whetted my appetite for a more detailed study of China's military modernisation. While at Stanford, Professor Scot D. Sagan, Professor John Wilson Lewis and Xue Litai were of immense help in making my short sojourn, a memorable one. Needless to add that without the personal efforts of Air Cmde Jasjit Singh, the necessary permission and clearances from the Indian government would have taken months to materialise.

Following my retirement from the IAF on 31 August 2002, the United Service Institution of India (USI), New Delhi, not only gave me the opportunity to pursue my study of China by awarding me a one-year fellowship, the Air Marshal Subroto Mukherjee Chair, but also made it possible for me to visit China as part of a four-member USI delegation in September 2002. All this was possible only with the generous help and guidance from Lieutenant General Satish Nambiar, (Retd) Director USI, and Major General YK Gera (Retd) Deputy. Director USI, Air Marshal Vir Narain (Retd) Chairman, Board of Management USI Centre for Research and Colonel VK Singh (Retd) Deputy Director (Administration). The librarian and staff at the USI were of immense help. I would also like to thank Shri Yashpal Bisht for his help in compiling this report.

I am also grateful to Air Marshal Bharat Kumar (Retd), Rear Admiral KR Menon (Retd), Major General VK Srivastava (Retd), Air Vice Marshal Kapil Kak (Retd) and many others for their encouragement and moral support when my spirits tended to flag. There are many other friends and colleagues in and outside the IAF from whom I learnt much but due to paucity of space, I cannot name all of them here.

Finally, I owe a great debt of gratitude to my wife Dr. PS Phadke, herself an author of four books, for putting up with my moods and for sending me back to my computer whenever I lost patience.

Ramesh Phadke

Introduction

By virtue of its enormous size, population and a three thousand year long civilisational history, China has always been an important country even when it was relatively weak. Its recent achievements in the economic, political and military fields have further enhanced its importance. For over a quarter century since 1978, China's rising power has been exercising the minds of strategists and defence analysts the world over. Its spectacular performance on the economic front and the rapid modernisation of its armed forces in its quest for what it calls, "Comprehensive National Power" (CNP), have cast a long shadow on the Asian strategic scene. It had quadrupled its 1978 Gross Domestic Product (GDP) by 1998. In Purchasing Power Parity (PPP) terms, China's GDP is second only to that of the US. Even after major reductions in the last two decades, China's armed forces are still the second largest in the world. A consistent, high single digit GDP growth would make significant funding available for its military modernisation. Since the interests of many major powers such as the US, Japan, and Russia converge in the Asia Pacific Region, this dramatic increase in its national power is even more significant for the strategic and economic future of the entire region. Other regional players watch with baited breath, China's march towards its goal of becoming a global power as it reshapes the regional power balance. No one quite knows the likely attitude and behaviour of an economically strong and militarily powerful China. Some believe that it may become even more assertive and aggressive while many others think it could well be a benign and even inward looking China. Then there is the third school that feels that China may implode under the weight of its many internal contradictions. What turn it finally takes is still in the realm of conjecture.

China's attempts to comprehensively overhaul and modernise its armed forces, its emphasis on access to modern technology and its search for the so called 'magic weapon', indicate that unlike in the past, China's leadership is determined to build a solid, all round military capability to fight not just at the 'gates' but well beyond its immediate periphery. Along with building its economic strength, China is determined to build its power projection capabilities. When Deng Xiaoping launched his 'four modernisations' in 1978, the military was placed last on the list but according to one Western scholar, "China has achieved two generations of military modernisation in just over fourteen years"; an impressive feat by any account. The interesting aspect of this rapid reform, re-organisation and re-equipment programme of the People's Liberation Army (PLA) is that unlike in the past, the Peoples Republic of China (PRC) is determined to acquire the necessary instrumentalities to fight and win a modern high-tech war.

The history of the last fifty-four years shows that while declaring its adherence to '*Panch Sheel*', the "Five Principles of Peaceful Coexistence" and a defensive military doctrine, China initiated no less than eight military conflicts. Most of these instances had to do with what China calls its vital national objective of reunification. To the outsider, however, it seems a clear case of Chinese expansionism. Having once made a notional territorial claim, China is often content to wait for many years until favourable conditions are obtained to translate that claim into reality. Coercion, and if that fails, compellence using military force have been the logical and natural next steps in China's strategy. Given this history and the marked propensity to use force, China's fast burgeoning military capabilities cannot be underestimated or ignored.

Recently, however, China's foreign policy conduct and style have become much more sophisticated. The new leadership is young and well versed in the art of diplomacy so that China's image is benign and more conformal with the contemporary international climate. This makes it even more difficult to gauge its intentions. In all probability, China's new rulers will take their country to higher levels of power, influence and prestige. Its initiative in the six country-negotiations to resolve the North Korean nuclear weapons question shows a new readiness to play a more active role in international affairs.

For India with which it shares, a 4,700-km-long disputed boundary, China's rising economic and military power poses a major and long term challenge, if not an extant military threat. For many millennia, the two large neighbours had coexisted in peace. China's occupation of Tibet in 1950 brought the newly independent Communist giant on India's doorstep. Even so, the mighty Himalayas kept them apart. China's avowed aim of national reunification includes the recovery of all claimed territories. This is worrisome to India as China claims the whole of Arunachal Pradesh, a state in the Indian Union. Further, China is not content with the status quo but is constantly expanding its activities in Pakistan, Myanmar, Bangladesh and Sri Lanka. So long as China makes inroads into South Asia by building strategic relationships with India's small neighbours, it would be difficult if not impossible for India to improve its relations with China. The memories of the 1962 Chinese invasion are still raw in many Indian minds. It is immaterial whether the Himalayan War resulted from mutual misunderstanding, miscalculation or misperception; the fact is that China used military force to "teach India a lesson" in 1962.

Despite a gradual improvement of relations and a very sizeable increase in bilateral trade, suspicions remain. It should be noted that in the recent past, China has resolved its border disputes with all its neighbours except India. China continues to remain in illegal possession of large chunks of Indian territory in Ladakh and in addition claims Arunachal Pradesh. This long outstanding border dispute, and China's consistent and continuous support to Pakistan in vastly improving the latter's strategic position vis-à-vis India are the two most serious impediments to peace and mutual trust. China's declarations of its desire to live in peace with its neighbours, thus, sound hollow. Given these ground realities, the rapid pace of China's military modernisation that would permit it to extend its influence and power well beyond its immediate proximity is a cause of serious concern for India. India cannot remain a mute witness while the power differential between the two neighbours continues to grow in China's favour.

There is a strong likelihood that by the end of the current decade, China would have acquired adequate military power to project it at considerable distances in the Asia-Pacific Region and that this would have serious implications for India's security. This study attempts to prove this hypothesis and hopes to improve our understanding of China's rapidly rising power and its possible implications for India' security.

Chapter 1 briefly examines China's grand strategy as it has evolved since the founding of the PRC in 1949 and the broad contours of its security behaviour. The chapter looks at the role of diplomacy and politics in the attainment of China's foreign policy goals.

Chapter 2 then goes on to discuss instances and patterns of China's use of force in different situations/conflicts including in the 1962 attack on India. China's coercive actions in what it called missile exercises across the Taiwan Strait in 1995 and again in 1996 are scrutinised to see if China's security approaches have changed in recent times.

Chapter 3 attempts to give a broad overview of the evolution of China's military doctrine. Mao exerted a strong and enduring influence on China's military thinking and strategy. He enshrined the "People's War" doctrine in his country's military/strategic lexicon until it became a dogma; and yet there is evidence to suggest that he was quite conscious of the role of technology and was already devising ways of modernising China's military in the 1950s but could not do so simply because of paucity of resources. The hallowed concept had however, started to give way to what came to be known as the doctrine of 'People's War Under Modern Conditions' during Mao's life time. The remarkable vision and foresight of Deng Xiaoping in boldly announcing that the threat of a major, nuclear or global war had abated at the time when the second Cold War was raging is also examined in so far as its true meaning and influence on China's military doctrine. The chapter then goes on to discuss the current doctrinal debate between the protagonists of the local war doctrine and those of the Revolution in Military Affairs (RMA) School.

Having discussed China's defence potential, in Chapters 4, 5, 6 & 7, I examine in some detail the process of China's military modernisation. The People's Liberation Army (PLA) played a crucial role during the long civil war and brought about a successful revolution. There is more to its name than mere symbolism. It has been the backbone of the State and the Party, and has been used as the primary instrument of national power by the PRC leadership. Even today the PLA owes its allegiance and loyalty to the Party and through it to the State. In recent times, the PLA might appear to have lost some of its importance and pre-eminence in the Chinese central decision making process but it continues to remain a vital organ of the State and a bulwark against domestic instability. In fact, due mainly to widespread reforms and regularisation the PLA has become more professinal and less political but has not lost its clout. In China, the other two wings of the national armed forces are known as PLA Navy (PLAN) and PLA Air Force (PLAAF). It is little wonder then that under this dispensation both the navy and the air force have never really enjoyed the freedom and autonomy that they deserve. China's Strategic Rocket Force or the Second Artillery, on the other hand, has always been regarded as the crème de la crème of the PLA. This wing of the armed forces continued to get full funding, protection and encourgement even during the ten-year long cultural revolution when the PLAN and PLAAF were not only neglected but suffered major disruptions and purges.

The Chinese leadership has always laid great stress on self-reliance in defence. Within months of the proclamation of the PRC, China signed the historic 14 February 1950 Treaty of Friendship with the then Soviet Union, cementing a relationship that was to prove exceptionally fruitful to the PRC for the next ten years. China soon developed a massive defence industry with Soviet assistance. In the last decade or so, China has reinvigorated its efforts to upgrade its military technologies. Indigenisation and foreign imports have both contributed to rapidly building China's aerospace, maritime, information warfare (IW), electronic warfare (EW), and other cutting-edge technology industries. China's younger generation is highly techno-savvy and this has affected the armed forces in a big way. Imported communications systems, navigation and reconnaissance satellites, super computing, lasers, simulation and artificial intelligence are today the most sought after areas of high technology which in due course of time would be assimilated, operationalised and integrated in China's existing high-tech assets with the sole purpose of putting China in the very forefront of global RMA powers. Chapter 8 traces the origins and development in this vital field of China's national endeavour, the many impediments to its progress, and the successes of self- reliance after the historic break with the Soviet Union. The contradictory demands of the supreme leadership during the Great Proletarian Cultural Revolution, the consistent and ongoing search for dual-use technologies from foreign sources and the prospects for China's indigenous defence capacity in the design, development, and production fields are also discussed in this chapter.

Chapter 9 is focused on the likely shape of China's overall military capabilities and preparedness in order to see their contribution in the field of power extension and power projection capabilities. The implications for Indian security; are examined in this chapter. This chapter sums up the major findings of the study and offers some possible future scenarios, should conflict erupt between India and China. The chapter also highlights the urgent need for India to develop the necessary expertise to understand and interpret China's rapidly developing national power and its likely employment in the region. As the name suggests, chapter 10 looks at the prospects of peace in the region in the backdrop of the study.

Here, it is essential to insert a word of caution. The PRC leadership is undoubtedly determined to enhance China's military power but its record of progress in the last decade has been uneven with some major achievements and, some failures. Exaggerating the 'China Threat' is counterproductive lest it became a self-fulfilling prophecy. Knowing the Chinese system is far from simple but one thing is certain: the PRC leadership may be secretly pleased if it finds that some of its major neighbours are indeed deterred by its military power. India would be playing into its hands if it allowed itself to fall in this trap. The PRC leadership desires that the world respect China – even if it is out of fear. It is, therefore, axiomatic that we guard against the tendency to overestimate the threat while assessing its potential.

The Bibliography placed at the end of the report lists all the major books and journals and other reference material used. I have also made use of the internet especially FBIS, BBC's SWB, and other sites including US Department of Defence, Stratfor.com and IISS.

1

Understanding China's Grand Strategy

"The Chinese have always been a great, courageous and industrious nation; it is only in modern times that they have fallen behind. And that was due entirely to oppression and exploitation by foreign imperialism and domestic reactionary governments. . . .Ours will no longer be a nation subject to insult and humiliation. We have stood up."

Mao Zedong, 21 September 1949[1]

"To continue to propel the modernization drive, to achieve the national reunification of the motherland, and to safeguard world peace and promote common development are the three historical tasks of the Chinese people in the new century."

China's National Defence White Paper-2002[2]

Introduction

The PRC came into being at the culmination of a nearly 150-year-long period of continuous turmoil and struggle against exploitation by foreign powers, domestic disruptions and factional fighting between different political forces. By the beginning of the nineteenth century, the Qing dynasty had already started showing signs of mortal decay. Historically, the

Chinese society has been kept together only when a strong central power ruled the country. So long as the emperor wielded adequate authority and demonstrated adequate virtuosity and ensured that the well being of his subjects was uppermost in his mind, he could hope to exercise a reasonable degree of control to maintain domestic harmony and order. In the Sinic system, the emperor was seen as having the mandate of heaven to rule the so-called Middle Kingdom. However, once he had exhausted this heavenly mandate, the people were theoretically free to search for an alternative.

The idea and concept of the Middle Kingdom comprising what is now east-central China, surrounded by barbarians along its vast frontiers and periphery meant that the emperor could effectively ignore the activities in the outlying areas so long as these did not directly impinge on the security of the Middle Kingdom and consequently that of the emperor himself. Throughout its long history, however, China was subjected to a series of invasions mostly from the north and the northeast, which every Chinese emperor had to fight or at least resist. Some of the emperors were themselves invaders, like the Mongols and the Manchu. Resisting invasions, however, was a costly drain on imperial resources and usually it was the cost of these wars that weakened the empire and made it susceptible to the next invasion.

The critical need for a strong and widely accepted central ruler who could keep external threats at bay without overly taxing the state is thus the central theme of Chinese history. A weak and divided Chinese society was usually ripe for foreign exploitation and hence internal instability and discord are always linked to external threats. European powers made full use of this fact to exploit China.

Since the Middle Kingdom was surrounded by 'barbarians' the Sinic hierarchical order did not recognise sovereign equality between the Empire and the surrounding states or kingdoms and thus only a tributary relationship existed in which the smaller states were considered inferior. In fact, until about the nineteenth century, China did not have a Ministry of Foreign Affairs, a 'barbarian handling section' served the purpose. Interaction between the people and the states was however, often based on Confucianism with its high ethical principles of justice, fair play and a general preference for non-violent means for conflict resolution. Despite this, the Chinese did not desist from the use of military force but often justified it as self- defence.

The 1901 Boxer Rebellion staged by the local people opposed to the continued exploitation of the country by the five foreign powers in collusion with the Chinese officialdom, marked the decline of the last empire. Debilitated by continuous internal disruptions and foreign interference, the Qing dynasty finally collapsed in 1911. Dr. Sun Yat Sen, a foreign educated Chinese, established the new Republic but soon real power slipped into the hands of a general, signalling a series of students protests, social and political turmoil, and warlordism that finally led to the forty-year-long civil war.

Although it was a great victory for Mao and his cohorts, the end of the forty-year long civil war, and the defeat of the Nationalists (Guomindang) did not mean voluntary acceptance of Communist leadership by the larger masses. According to Ranganathan, a former Indian ambassador to China, it was more out of the disgust for the Nationalists and their ineffective resistance to the Japanese that the Chinese people chose the Communists. "The appeal to Chinese patriotism which brought the Party to power did not automatically win adherents to the possibility of a Marxist-Leninist framework of governance", he adds. Further, local insurgencies, subversion and banditry were common. Some remnants of Guomindang forces continued to wage war in the south west of the country and in Burma. Communism was not popular with the urban middle classes. They were hoping for "more freedom rather than the discipline which the Chinese Communist Party hoped to impose."[3] The Nationalists had been defeated but not eliminated. Having fled to Formosa (Taiwan) just across the Taiwan Straits, they were busy mounting armed attacks against the Mainland, with American backing. It was thus natural for the PRC leadership to develop a high degree of xenophobia. China has remained a paranoid power ever since.

International Relations: Politics and Dimplomacy

The external environment was no less inimical with the Americans supporting Chiang Kai-shek's forces on Taiwan, and the threat of war looming large on the Korean Peninsula. Mao's first major move was to find a dependable friend. He tried to build a strategic and ideologically strengthened relationship with the USSR. There was already some sympathy among the Chinese people for the Soviet Union. "In July 1919, the Soviet Union announced that it would voluntarily and unilaterally withdraw from the territories acquired by the Czars; renounce the Boxer indemnities; and return without compensation, the Chinese Eastern Railway and other concessions. This announcement had a powerful pro-Soviet impact on the radical Chinese intellectuals."[4] It paved the way for a full flowering of friendship with the newly emerging Communist Party in China in the light of which Mao's 'leaning' towards the USSR was only to be expected. With the signing of the 14 February 1950 Treaty of Friendship with the USSR, there began to flow a huge amount of technical, military and economic assistance in the form of aircraft, tanks, guns and Russian engineers, technicians and specialists. Mao then began to build China's economy and industry on the Soviet model. Sino-Soviet friendship was thus to become the bedrock of Chinese development and progress.

As far back as in 1937, Mao had planned to incorporate the two outlying regions of Xinjiang and Tibet as autonomous provinces of the new People's Republic. He realised that the

PRC could not treat the security of its frontiers lightly as during the imperial period. Seeing the threat of foreign exploitation in Tibet where a feudal system still ruled, in March 1950, Mao sent the PLA to liberate it "from the oppression of feudal orthodoxy and blind faith". Tibet had historically been home to many anti-Chinese revolts and had never accepted Chinese sovereignty. Its political status frequently depended on the strength or weakness of the central government in Beijing and the interest other powers, notably Russia and Great Britain, took in this inaccessible and often inconsequential trans-Himalayan kingdom. So long as Tibet continued as a buffer between Czarist Russia and British India, the governments in Delhi and London were quite contented. Mao was, however, not happy with this arrangement. It was dangerous to leave an outlying region such as Tibet exposed to the machinations of powerful neighbours. The recently independent India might entertain imperial ideas and exercise undesirable influence on Tibet. Hoping for support from India and the USA, the Tibetan leadership employed delaying tactics for a while when Beijing increased pressure in 1950, but capitulated without much resistance. Having secured the southwestern flank, Mao turned his attention to dealing with the threat on Korean Peninsula.

The prompt American response to the North Korean invasion of South Korea in June 1950 was seen as a direct threat to China. With the rapid advance of the UN forces to the Yalu River that formed China's southern boundary with North Korea, it became increasingly difficult for China to stay out of the conflict. President Truman's decision to send the American Seventh Fleet to patrol the Taiwan Straits as a precaution against possible Chinese attempts to take Taiwan by force, added a new dimension to the threat. Despite US assurances that China was not its target, the Chinese Communist Party (CCP) decided to send its forces, the so-called "Chinese People's Volunteers", into North Korea. The nearly three-year long Korean War, ended in a fragile armistice with the opposing forces at about the same geographical location from where they had started. The Chinese fought the war with Soviet political support and weapons, further strengthening the Sino-Soviet relationship against a common enemy. "The Sino-Soviet alliance was thus the mainstay of Chinese foreign policy for about a decade"[5]

The Americans continued to back the Nationalists on Taiwan. This resulted in a number of skirmishes when the Taiwanese resisted Chinese attempts to capture some of the small islands in the straits. This further strengthened Mao's resolve to oppose Western imperialism. His attempts to export Communist ideology to other countries that were fighting the forces of imperialism and colonialism became part of this overall fight. There was, thus, a substantive ideological content to the PRC's foreign policy during the Mao era.

The Mao Era

When the threat of superpower nuclear confrontation became clear, Mao did not want to be left behind; he was fully conscious of the power and status that went with the possession nuclear weapons. Although he publicly dismissed the threat of nuclear weapons as hollow (he called them "paper tigers"), he, nevertheless, gave the green signal to build China's own atomic bomb in 1955. It was this decision and Mao's insistence that the USSR give China a prototype of an atomic bomb that were to sour relations between the two Communist giants in the late 1950s.

Although the Russians eventually did not provide China with the bomb design, the economic and technological assistance that they gave China was to play a crucial role in building China's strategic weapons and its industrial capacity. Mao, however, was soon to be disenchanted with what he saw as Russian attempts to play the big brother and leader of the world Communist movement, especially after the death of Stalin.

There are many reasons ascribed to the Sino-Soviet split that became public in 1963. According one scholar, "Differences in ideological and strategic outlook accounted for the Sino-Soviet rift". By about the end of the 1950s, the Soviet leadership was worried about the threat of a nuclear conflict with the West. Khrushchev then advocated peaceful coexistence with the United States. The Soviets argued that war was no longer inevitable and that the transition to socialism could be peaceful. Mao opposed this "revisionist" line and argued for a more militant worldwide struggle against capitalism. Mao considered himself as the rightful leader of the global Communist movement. He thought that the softer approach of the USSR would compromise the Chinese Communist regime as well as its national goals like the recovery of Taiwan".[6] Lack of Soviet support to the PRC during the 1959 Taiwan Straits crisis and the 1962 border dispute with India, were also seen as examples of Soviet betrayal. The assiduously built and long cherished Sino-Soviet alliance was crumbling. For the next decade, China found itself internationally isolated. The PRC leadership, however, had made contingency plans for just such an eventuality. Mao now saw a nuclear threat developing from the Soviet Union.

China's problems with India began almost immediately after the founding of the PRC although these did not come into the public domain for some time. India was the second non-Communist country to extend its recognition to the Communist regime in China, but soon thereafter, the Chinese military invasion of Tibet was to raise the spectre of the Chinese threat in Indian minds. India tried to resolve the Tibet issue diplomatically. In a diplomatic note sent on 21 October 1950, India told the PRC, "China's occupation of Tibet would not be viewed kindly by those opposed to China's admission to the United Nations and the Security Council. This was interpreted by Beijing as a hint that India would oppose China's entry into the United Nations unless it halted its military move into Tibet". In its reply, it said, "Tibet was in integral part of Chinese territory and the problem of Tibet is entirely a domestic problem of China". China also warned India that if it used China's move into Tibet as a pretext to obstruct China's entry into the UN, it would be seen as a demonstration of the unfriendly and hostile attitude having been affected by foreign influences hostile to China in Tibet". India sent another note on 01 November 1950 directly raising the question of Tibet's status and suggesting Tibetan autonomy within the framework of Chinese suzerainty. "Beijing replied on November 16, demanding, in effect, that India choose between accepting Beijing's unlimited sovereignty over Tibet or confrontation with China."[7] China had once again rebuffed Indian attempts to uphold to the degree possible, the freedom, independence or autonomy of Tibet.

Nehru was obviously aware of China's potential power, the Communist regime's proclivity to use force and the complexity of the Tibetan status. It was perhaps because he was too keenly conscious of a rising Communist power in India's immediate neighbourhood that he tried to neutralise the emerging threat by building friendly relations with China. He made it clear that India had no designs in Tibet and was ready to accept China's sovereignty over the region provided, of course, that the PRC gave due consideration to the emotional and religious bonds that had historically existed between the people of India and Tibet. By this, he meant freedom for the Tibetans to practise their religion without undue interference from Beijing and Indian access to places in Tibet that it had traditionally enjoyed. The futility of trying to influence the strategic decision of a strong neighbour without the backing of adequate military force should have dawned on India. India continued to entertain a kind of romanticism about its image as a country interested in world peace and of its moral influence on others. Little did India realise that it was dealing with a country that had recently emerged from a forty year-long civil war and had a leadership that was hardened to the reality of war and power.

India's attempt to salvage a semblance of autonomy for Tibet was seen as blatant interference in China's domestic affairs. China had perfunctorily dismissed Tibetan and Indian objections to its grand objective of national reunification. India was concerned that China would soon populate the region with the majority Han Chinese people thereby gradually but completely extinguishing Tibetan identity and distinctive culture, which China saw as contrary to its Communist ideology and a potent threat to the regime. China did not want to allow two different political systems to coexist because of the fear that other minorities or nationalities might also demand the same kind of freedoms and weaken the Communist Party's hold over the country.

India, on the other hand, could not countenance the militarisation of a hitherto peaceful region. To be sure, some Indian leaders had warned Nehru of the potential for trouble on India's northern borders unless the border was demarcated to the mutual satisfaction of the two neighbours. Foremost among them was Sardar Patel, the then Home Minister and Deputy Prime Minister of India. He had just concluded the most complex and complicated exercise of integrating some 562 former princely states into the newly formed Indian Union and was acutely aware of the problems left over from the British colonial era. He did not want the emergence of another potential military and political threat on India's vulnerable northern, northwestern and eastern borders when India's economy and military power was woefully inadequate to deal with it. He also foresaw a double threat emanating from China and Pakistan although at that time, China had not yet shown any inclination to befriend the latter. Patel's warning was to prove prophetic when during the 1965 Indo-Pak conflict, China made open threats to India.

Soon after the Chinese occupation of Tibet, in an effort to improve relations, India signed a trade agreement with China and relinquished the rights and privileges that it had inherited from the British. Soon thereafter, there arose the issue of Chinese maps showing large tracts of Indian territory under Chinese control. When Nehru pointed this out to Zhou Enlai, the latter assured him that the maps were old and the new editions would be revised to correct the error. This was, however, a duplicitous answer, as China had no intention to relinquish its claims and was busy planning the construction of a highway linking Tibet with Xinjiang that was to eventually pass through Indian territory. Relations soured despite many attempts to start negotiations. China finally invaded India in response to India's feeble attempts to 'show the flag' by implementing the much maligned 'forward policy'. The PRC, however, succeeded in teaching India a lesson and effectively severing all contacts for over a decade until they were restored in 1976. Mao had thus effectively neutralised India and secured a region that had historically been well outside China's control. In the bargain, China had also caused India deep hurt and lost forever the trust of a useful and peaceful neighbour. Even today, despite many attempts by India to place the Sino-Indian relationship on a firm foundation of mutual confidence and trust, there remain many of the same fears and doubts in Indian minds. It is, therefore, difficult to say if the 'teaching of lesson' to India indeed assisted China's overall foreign policy objectives.

By about 1962-63, the Sino-Soviet rupture was near complete but China continued to receive Russian assistance well after the departure of the Russian specialists, engineers, and technicians in 1960. Mao now began to berate the Soviet leadership for abandoning the goal of spreading socialism across the world. Stung by Russia's neutrality during the Sino-Indian conflict and Khrushchev's détente with the US, the Chinese leadership now embarked on a programme to assist

Third World countries to garner their support. Exports of Russian origin arms and equipment and support to local insurgencies were the most frequently and commonly used tools. The Soviet invasion of Czechoslovakia in 1968 and the Brezhnev doctrine of 'limited sovereignty' rang alarm bells in China. From then on, Mao thought that a nuclear war was not only possible but was imminent. That the threat of a Soviet attack was real at the time is confirmed by Mao's decision to set up the 'third front', by shifting China's strategic industry to the deep hinterland of western China to protect it from such a Russian nuclear attack. China was now more paranoid than ever before.

Soon thereafter began China's search for a new ally to counter the Soviet Union. China tempered its anti-imperial rhetoric. Having exploded its first atomic bomb in 1964, China quickly developed its missile forces and by 1967, became a thermonuclear power. The threat of nuclear blackmail was, thus, taken care of but China was still internationally isolated. It is at this time that its ally, Pakistan, came to its rescue.

Secret parleys had begun between China and the US with Pakistan as the intermediary. These later resulted in the now famous ping-pong diplomacy and the secret visit of Henry Kissinger, the US National Security Advisor, to China in July 1971. That was the time when tensions on the Indian subcontinent were mounting with Pakistani repression and genocide of the Bengalis demanding freedom in East Pakistan. India and Pakistan were once again on the brink of war. While China did not militarily threaten India, it provided consistent political support to Pakistan. Its foreign minister said in the UN that India's claims of the refugee influx were untenable. "There were thousands of Tibetan refugees living in India. Would India launch an attack on Tibet using that as the pretext?"[8] So beholden were Nixon and Zhou Enlai to Yahya Khan, the then President of Pakistan, that the US sent its Seventh Fleet to the Bay of Bengal to pressurise India into an early ceasefire.[9] Talking about it during their historic meeting in Beijing in February 1972, both Nixon and Zhou sounded extremely contemptuous of India. The now declassified transcripts of their conversations clearly show Chinese visceral dislike of India, Nehru and, more importantly, the fear of Indian influence spreading in the Southeast Asian region. Moreover, to think that all this was based on Nehru's book, *Discovery of India* in which he had highlighted the sway of Indian culture in far off lands further shows Chinese fears of India's cultural influence.[10] The 1971 Indo-Soviet Friendship Treaty also raised security concerns in China, which began to see Soviet support to India and Vietnam as an attempt at its strategic encirclement.

Despite its animosity towards the Western bloc, China was quick to realise the importance of Pakistan, which in the mid-Fifties was already an important member in the Central Treaty Organistaion (CENTO) and Southeast Asian Treaty Organisation (SEATO) alliances when Zhow Enlai, the Chinese premier met the then Pakistani Prime Minister Mohammed Ali Bogra at the 1955 Bandung Conference of Afro-Asian countries, he categorically told him that China saw no possibility of their respective countries' interests ever clashing and assured him that Pakistan's friendship with the USA would not be allowed to come in the way of building a lasting friendly relationship to their mutual advantage. This not only shows the strong element of pragmatism in China's foreign policy approach but also the generally well-known Chinese fondness for what its traditional classics call, using the "borrowed sword". Pakistan played a stellar role in arranging a secret meeting between Henry Kissinger and Zhou Enlai that finally resulted in the historic Nixon visit to Beijing in February 1972. Unlike in the early 1950s when the Soviets had taken the initiative in co-opting China in the global fight against capitalism, this time it was the US that was in need of a partner to oppose the Soviets. Soon after the 1962 border war with India, China signed an agreement with Pakistan under which it received on lease a sizeable piece of Indian territory in Northern Jammu and Kashmir. Pakistan now became China's 'cat's paw' in constraining India. The lease further complicated the Sino-Indian border dispute.

China continues to provide unstinting political and material support to Pakistan. During the 1965 Indo-Pakistan conflict, China threatened hostilities on the northern front. Its support to Pakistan during the 1971 War was limited to the political level but China pragmatically began to build its relations with Bangladesh soon after the assassination of Sheikh Mujibur Rehman. Friendship with Pakistan was to prove an important achievement of the Mao era.

China's relations with its other neighbours were not exactly easy. The memories of the Japanese invasion were still fresh. In addition, following the end of World War II, Japan was seen as an American surrogate and, hence, a potential threat. North Korea was for a long time a reliable partner in spreading Communism across the world and a useful buffer between China and South Korea where the Americans maintained a large military presence and for a long time also placed nuclear weapons.

China's relations with Vietnam have historically been quite tense. China had once held sway over large parts of that country. North Vietnam was given much military and political support during the latter's fight against the West in the 1960s, but China was never able to tame Vietnam into blindly acknowledging China's ideological and strategic supremacy in the region. A large ethnic Chinese population lived in the northern border areas of Vietnam. The Vietnam authorities always feared that the PRC might use this population to create trouble and once again establish Chinese influence, if not control, over Vietnam.

Burma, the other small neighbour, also faced somewhat similar problems. Its northern border areas with China were home to a number of insurgent groups that were actively supported by China and there was always the danger of things getting out of hand. Burma, therefore, pragmatically, though somewhat belatedly, saw the writing on the wall when the Sino-Indian border dispute became public in the late Fifties, and negotiated a border agreement with China in 1960. The Chinese leadership achieved a small coup by showing China's reasonableness to the world by readily relinquishing its claim over some small portions of Burmese territory, and more significantly, acknowledging the existence of the McMahon Line although the boundary was never officially referred to as that. Thereafter, Burma avoided taking sides but knew full well the PRC's ability to foment trouble by supporting the many insurgent groups and Communists in the north.

In 1971, China took its seat on the UN Security Council, now that the American opposition to China's entry to the world body had evaporated, and soon became quite comfortable as the member of the "Permanent Five". This was a major gain of the switch in its foreign policy. The PRC was now 'aligned' with the US against the Soviet Union, an erstwhile ally that had been instrumental in building China's industrial and military strength and helping the PRC to build nuclear weapon in a remarkably short time. The Sino-American friendship, however, took some time to flourish. China was keen to obtain modern technology from the US and other Western countries; mending fences with the Americans opened doors in many a Western country that had until then been very wary of Communist China. It is noteworthy that the PRC capitalised on this friendship with the US and extracted from the Americans their support for its "one-China policy". The Americans did not abandon Taiwan but said that a peaceful solution should be found to the Taiwan problem and that China should not resort to the use of force. China had, thus, secured its position on Taiwan.

The Sino-American honeymoon, however, lasted only until the advent of the Reagan Administration in the early 1980s. This was because of Reagan's oft-repeated attacks on Communism and his resolve to defeat the "evil empire" (the USSR). The PRC now realised that there were limits to its friendship with the US and that it was becoming increasingly difficult to conduct an independent foreign policy. China did not want to play a subordinate role. Even so, China derived as much benefit as it could from this relationship.[11] China received much assistance in the field of high technology, including super computers, and this assistance continued well after the US imposed sanctions on the PRC following the June 1989 Tiananmen crackdown.

The Deng Era

With the passing away of Mao and Zhou in 1976, China's new leadership under Deng Xiaoping decided to make a radical change to its Maoist era foreign policy. The main aim of which was to open China to the world so that through economic reforms it could build its national power. Wu Xinbo says, "Beginning in the early 1980s, several developments led to the reappraisal of security in the PRC. Foremost among them was the high priority accorded to national development. During the Mao era, the first priority in domestic affairs had been to ensure regime security. To that end, Beijing had adopted the principle of "taking class struggle as the key link" and launched political campaigns to liquidate all sorts of class enemies". To be sure, Deng Xiaoping was also concerned with regime security, but did not see any serious danger to the leadership of the Communist Party. China's weakness as a nation, especially its backward economy was his main worry. "Deng also believed that the best way to promote regime security was through improvement of public welfare, not political suppression. For this reason, in the late 1970s, the CCP decided to adopt the 'four modernisations'-

modernisation of agriculture, industry, national defence and science and technology- as the highest priority on the national agenda." To promote modernisation, Beijing adopted a reform and open door policy. Deng still shared Mao's opinion that the international system was anarchic and chaotic. He differed with Mao on two points: he saw the international system as a source more of opportunity than danger, and he was more inclined toward cooperation than confrontation in pursuing national interests."[12]

Around this time, the PRC's immediate security environment was also improving. Deng boldly declared that the threat of a major or nuclear war had receded. This helped create a favourable external atmosphere for China's modernisation programme. Beijing decided to improve its relations with the Soviets by toning down its anti-Soviet rhetoric. In the mid-1980s, as China and the Soviet Union moved to improve bilateral relations, the possibility of a major conflict between them gradually receded. "For the first time since 1949, PRC leaders did not have to worry about direct external threats. As a result, the military element, both as a security concern and as an approach to security goals, was downplayed."[13]

Beijing has since then tried to present a more refined and benign face to the world. Once the economic reforms had begun to show positive results, Deng was able to tone down public rhetoric. Even so, the core goals of the PRC regime have remained the same, "economic development, national unification and opposition to hegemonism." With the opening up of the economy came opportunities for close and continuing contacts with the world that spread across all sections of the Chinese society such as the students, academia, scientists, military officers, businessmen, industrialists, members of the government, politicians and the bureaucracy. Economic decisions such as the creation of Special Export Zones (SEZ), Joint Ventures (JV), Village Level Enterprise (VLE) and a host of other initiatives also served China's foreign policy goals by creating a stake for the investors as well as for the domestic entrepreneurs to develop mutually advantageous relationships with many countries.

Briefly stated, China's 'revolutionary diplomacy' of the 1960s and 1970s when it supported world revolution and exported sizeable amounts of arms to countries that appeared to be antagonistic to the US led Western alliance, gradually gave way to economic and business diplomacy aimed at gaining access to new trade contacts, dual use technologies and investments. The role played by the overseas Chinese community in China's neighbourhood, especially in Hong Kong and Taiwan in helping China with the much-needed Foreign Direct Investment (FDI) also needs to be highlighted because without this support it, would have been difficult for the PRC to attract FDI from other countries. This shows that despite differences over ideology, the overseas Chinese mainly of Hongkong, did not hesitate to invest in China the moment it opened up. This speaks volumes for the patriotic spirit and the strength of the deep bonds and links of the overseas Chinese with their motherland, and the statesmanship and diplomatic skills of the PRC leadership in assuring them that their investments were safe and that they were needed for the larger goal of building China's economic power.

Of late, China's diplomacy has become even more proactive and sophisticated. Gone are the days when dowdy men inhabited the corridors of power. Today, the Chinese diplomats are suave, personable, cultured, and well versed in the ways of the modern globalised world and are at home with English and many other European languages. They very quietly but firmly convey the growing importance of the PRC in global affairs. In short, China's international face shows poise, pride and pragmatism without being petty, paranoid or overbearing.

China's military diplomacy has also been very active in the past decade. According to the 2002 White Paper on Defence, China has posted military attachés in over one hundred countries.[14] The comprehensive training programmes for the PLA officers that had begun in the late 1970s and early 1980s in the US have been going on without any disruption with many PLA officers working for their doctoral degrees in the US and other Western universities. The PRC military officers were at one time trained in international relations by some well known names in the US political and academic circles.

Kenneth W. Allen and Eric A. McVadon, long time China watchers, in a report on China's Foreign Military Relations say, "The PLA has tailored its foreign relations programme to maximise benefits for itself, while minimising the amount of information the PLA provides to other countries."[15] The aim of developing and strengthening these military ties is to shape the international security environment, gain access to new ideas and technologies, and build bonds of friendship and help the military modernisation programme of the PLA.

Present Scenario

If Deng was successful in placing China firmly on the international map, his successors have done even better. Jiang Zemin, was chosen by Deng to succeed him. He continued with the reform policy of his predecessor. He is rightly credited with the peaceful return of Hong Kong and Macau to China and the maintenance of good relations with the US despite some serious setbacks. The two most notable incidents being the tension following the 'accidental' bombing of the Chinese embassy in Belgrade during the 1999 Kosovo bombing campaign and the April 2001 collision between a Chinese fighter and an American Spy-plane gathering electronic intelligence along China's southern coast. In both cases, the Chinese demonstrated their anger and displeasure but stopped short of causing any permanent damage to this all-important relationship. Jiang Zemin also succeeded in ensuring an orderly and peaceful change of guard when Hu Jintao and other relatively young leaders took over power, a unique achievement since all leadership changes in the past were problematic.

China has also been successful in gradually strengthening its relations with Russia to the extent that a former ally turned adversary is today a strategic partner. Sino-Soviet relations had begun to improve after Gorbachev's visit to Vladivostok in July 1986 when he announced that the two countries could live in peace. Starting in 1987, he ordered sizeable troop reductions along the border with China and a joint Soviet-Chinese commission began work on the resolution of the border dispute.[16] After the collapse of the Soviet Union and the consequent decline in Russia's power and status, the two countries could treat each other as equals. Russia's failing economy soon made it imperative for it to generate revenues from arms exports and here China was to prove a very lucrative market. By 1992, the two had signed the historic deal for the sale of the Su-27 air-superiority fighter that had not been exported until then. Both Russia and China need each other, the former depends on the burgeoning arms sales relationship while the latter hopes to obtain high-technology weapon systems. This mutually advantageous equation is, however, not without the usual vicissitudes of international relations. First, Russia is worried that its fast dwindling population makes it difficult for it to safeguard its interests in the far east of the country, as fewer Russians are willing to work in these remote areas. Chinese presence in cross border trade is already sizeable and an increasing number of Chinese are swamping the region with their goods and providing much-needed employment to the poor Russians. According to an American scholar, Dr Jeffrey Record, there is the fear of China absorbing the Russian Far East (RFE). "Moscow's control over the RFE has steadily weakened since the Soviet Union's demise; the RFE's economy is fast becoming a subsidiary of China's; and Chinese demographic infiltration of the RFE could eventually raise the issue of the RFE's self-determination in China's favour." He goes on to add, "The post-Cold War rapprochement between China and Russia has not eliminated centuries-old national and racial animosities between the two countries, animosities that can only be heightened by the growth of Chinese economic influence and demographic 'aggression' in the RFE."[17] Some Russians also point out the potential for Chinese expansionism in this area especially since there is very little that Russia can do to check the growing Chinese influence. While the Sino-Russian border is no doubt tranquil, there remains the potential for unresolved border issues to escalate in the future. A Russian scholar, however, strongly feels that it is in Russia's long-term interest to develop an enduring relationship with the PRC. He also cautions against succumbing to the imaginary fears of the 'yellow peril'. He recommends that Russia bring about an orderly development of the RFE to attract multinational investment and employment, and make more vigorous attempts to improve trade with China.[18]

Russia has not so far given the very latest, cutting-edge military technology to China but that will happen sooner than later. The PRC knows that Russia's concerns for an eastward expanding North Atlantic Treaty Organisation (NATO) would force it to maintain a relatively light presence in the East and at the same time increase its dependence on China as a partner for collective bargaining with the West. Sino-Russian ties are likely to remain strong now that the US presence in Central Asia and Iraq is unlikely to reduce in the near future. The PRC, on the other hand, knows that it would take Russia at least two decades to enhance its overall power status and hence its presence in the Asia-Pacific Region is likely to remain at a low key. The arms relationship between the two will most likely endure to their mutual advantage.

Many PRC scholars have noted the incipient nationalist stirrings in Japan and are convinced that sooner than later Japan, its archenemy for the better part of the last century, would again pose a threat once it embarks on enhancing its military strength. For the moment, the Chinese are only obliquely indicating the fears of a rising militaristic Japan. So long as Japan's security is tied to the US through the Security Treaty, the Chinese feel safe and, hence, despite the anti-American public rhetoric, the PRC has a stake in the US maintaining a sizeable and visible military presence in the Asia-Pacific region thereby removing the motive for Japan's militarisation. The PRC has also built a very important economic and trade relationship with Japan with China accounting for one fifth of Japan's imports. It is, therefore, in China's interest to try its utmost to maintain a peaceful environment in its immediate proximity without allowing Japan to become too powerful.

For many decades, the PRC leadership preferred bilateral interaction with other states in the belief that China stood a better chance of persuading the other side whereas in a multi-lateral arrangement, the smaller countries could gang up against it. This has been changing to some extent in recent years with the PRC, in fact, encouraging multilateral forums such as Association of Southeast Asian Nations (ASEAN), ASEAN Regional Forum (ARF) and Asia-Pacific Cooperation (APC) in the hope that China would, by joining these forums, be better placed to assuage the fears of the smaller players in the region. While China no doubt has been steadily improving its relations with its South China Sea neighbours, concerns about China's possible expansion, especially for the control of the Spratly and Paracels Islands and the under-sea natural resources remain. With China fast replacing the former booming economies of the region as a preferred destination for FDI, there is a growing fear that this rising giant would soon swamp their economies. A majority of the South China Sea littoral states, in fact, fear China, for they do not know how it would behave when its military modernisation programme reaches fruition in the near future. These countries would, therefore, like the US to maintain a robust presence in the region. Post 9/11 the global war on terror has given the US a greater justification for its presence in the region and this is causing some worries in Beijing.

The PRC has also been active in building its relations with the South Asian countries. It has all along been a major arms donor/exporter to Pakistan, Myanmar, Sri Lanka and Bangladesh. With the signing of the December 2002 defence treaty with Bangladesh, it has further cemented its friendship with that country. Its commitment to upgrade the port facilities at Gwadar on Pakistan's Makran coast and the building of the 'friendship highway' through Pakistan have given China a strong foothold in that country and further reinforced its 'all-weather' relationship. China's relations with Myanmar continue to remain close with the latter depending on China for its armament needs. China has also built airfields, ports and many other facilities in Myanmar which it can readily use in an emergency. Even so, China has few genuine friends. India's small neighbours and North Korea are the only countries that are perhaps not concerned by China's quest for power projection capabilities. The smaller South Asian countries look to China to balance India. "Implicit in all of this is the unspoken awareness that China maintains a large and powerful military presence on India's Northern borders and has fought one war against India and several other wars on behalf of smaller neighbouring ally states". (Korea-1950-53, Vietnam 1965-68, Kampuchea 1979)[19]

Assessment

It is not easy to assess China's intentions and true power potential. At one level, it is like any developing country with widespread poverty, underdevelopment and unemployment in the hinterland. Yet its GDP is the second largest in the world in PPP terms. Even so, there are some serious problems. "In terms of arable land per capita, China ranks twelfth from the bottom among all nations of the world. With 0.08 hectares per capita compared to a world average of 0.24, China falls just above Bangladesh and Egypt and on the same level as Haiti, the Philippines, and Vietnam."[20] This means that there are limits to which China can increase its agricultural output. Another problem is the highly unequal distribution of its population. The population of its far-flung provinces of Xinjiang and Tibet is extremely sparse compared to the eastern and coastal regions. China's efforts to Sinicise these distant provinces have resulted in the demographic balance fast swinging in favour of the majority Han Chinese population. This has brought unhappiness and fear of majority domination to the local populace in these regions.

There is a consensus among China watchers like Michael Swaine and Ashley Tellis,[21] and David Shambaugh[22] and many others that China follows a calculative strategy in its international relations. In my opinion, its quest for CNP is aimed at the reestablishment of the Chinese world order. China is convinced that without first developing its economic and military power it cannot achieve its foreign policy goals that are the safeguarding of China's sovereignty, national unity and domestic cohesion and internal stability. Swaine and Tellis call it a 'calculative security strategy'. In the past, China has changed its approach to national security depending on its overall power position in the international environment. When strong, it followed an assertive approach; when weak, it showed a readiness to adapt to the situation. In their opinion, "The term 'calculative'..... is, defined in substantive terms as a pragmatic approach that emphasises the primacy of internal economic growth and stability, the nurturing of amicable international relations, the relative restraint in the use of force combined with increasing efforts to create a more modern military, and the continued search for asymmetric gains internationally."[23]

Writing in 1985, Gerald Segal came to the same conclusion while examining China's methods of use of force but he simply called it a flexible policy where pragmatism overshadowed every other consideration including ideology, tradition and culture.[24] To be sure, that was a different era under a revolutionary leadership. It is not necessary that China would continue to follow the same pattern in the future. To recapitulate, China's foreign policy objectives are:

- Maintain a peaceful international environment to gain time to build its strategic strength.
- Prevent a Unilateral Declaration of Independence (UDI) by Taiwan.
- Ensure regime survival.
- Continue its efforts at national reunification without causing too much alarm especially in South China Sea.
- Improve relations without giving up its territorial claims.
- Prevent the rise of a militarist Japan.
- Ensure economic and trade development.
- Build or improve relations with EU and countries seen as being antagonistic to the US and India.
- Conduct foreign policy to improve access to high technology.
- Slowly build military capabilities to extend influence and control over South China Sea and the approaches to the Indian Ocean to safeguard energy and trade routes.
- Ensure domestic cohesion and internal stability by reducing regional inequalities and improving the standard of living of its population.

- Accelerate the economic development of its far-flung frontier regions to keep them peaceful and to extract their cooperation in the overall national effort.
- Ensure overall development in Xinjiang and Tibet by sending to those regions qualified, motivated and loyal party workers who would keep a tight control while, at the same time, win the hearts and minds of the local people.
- Build a position of pre-eminence in the Asia Pacific Region in the short term and compete for global power in the long term.

The PRC leadership continues to follow a realist and *realpolitik* approach to international relations. The general concern, however, seems to be about the persistent Chinese tendency to work outside the generally accepted norms of international rules and behaviour. In fact, the *raison d'etre* of the PRC's grand strategy is to become strong and powerful enough to deter US hegemony and dictate international 'rule making'. What China wants is nothing short of an unconditional recognition by the world community of its territorial integrity, including its claims and special rights in the South China Sea and the acceptance by the lesser players of the PRC as the preponderant power in the Southeast Asian region. It is safe to assume that for the foreseeable future, while China is preoccupied with its quest for enhancing its CNP, it would maintain a working relationship with the US, without compromising its core interests such as national reunification. Managing its relations with the US is perhaps the most important foreign policy and indeed grand strategy objective of the PRC. China knows that it cannot afford to jeopardise its trade with the US. Chinese economy is driven by exports. Its large-volume, low-tech character makes China vulnerable to trade wars with other countries. The sizeable investments by US multinational corporations in China have already created a stake in its wellbeing. This was illustrated during a recent incident. A PLAAF J-8II fighter collided with an American EP-3 electronic intelligence gathering aircraft killing the Chinese pilot and forcing the damaged American aircraft to land on Hainan Island in April 2001. The incident was seen as escalating into a major diplomatic spat between the two countries. Such was the stake and power of American business groups with large investments in China that the US State Department telephones did not stop ringing until the Bush Administration toned down its criticism of the PRC.[25]

The US may well try to restrain China's rapid progress towards becoming its peer competitor but there are limits to this and both China and the US would try their utmost to maintain at least a working relationship with each other. China may also use American help to manage the situation on the Korean Peninsula, but, at the same time, it would continue to play the North Korean card to leverage the US on its support for Taiwan. China would also continue its policy of constraining India while at the same time improving trade and other relations to keep India from growing too distant. It is unlikely to tolerate a regional competitor and, hence, would always remind India of its subordinate position vis-à-vis the PRC. The need to maintain a peaceful environment is of critical importance while China builds its CNP. It would, therefore, try to avoid raising tensions in the region. The moment it attains the necessary power potential to challenge the extant global order it will definitely do so. It is for this reason that China is often called a "dissatisfied power".[26] India is the only country in the region that can impede China's grand strategy objectives and, hence, China's South Asia policy is unlikely to change as it is specially designed to maintain indirect pressure on India.

Notes

1 Quoted in Wu Xinbo, "China- Security Practice of a Modernizing and Ascending Power", in Muthiah Algappa (ed), *Asian Security Practice* (Stanford: Stanford University Press, 1998), p.115.

2 Xinhua Overview of White Paper on "China's National Defence in 2002", available on the internet at FBIS Document No. FBIS-CHI- 2002-1209 inserted on 09 December 2002.

3 CV Ranganathan & Vinod Khanna, *India and China: The Way Ahead After Mao's India War* (New Delhi: Har Anand Publications, 2000), p.63

4 J S Bajwa, *Modernisation of the PLA* (New Delhi: Lancer Publishers, 2002), p. 20.

5 Algappa, n 1.

6 Ibid., p. 93.

7 John W Garver, *Protracted Contest: Sino-Indian Rivalry in the Twentieth Century* (Seattle: University of Washington Press, 2001), p. 46.

8 Ibid.

9 For a deeper insight into China's view of Indian policies towards the situation in East Pakistan and China's determination to safeguard the interests of its friend, Pakistan, see "Record of Historic Richard Nixon-Zhou Enlai Talks in February 1972. Now Declassified", internet version at <http://www.gwu.edu/~nsarchiv/nsa/publications/DOC_readers/kissinger/nixzhou/index.html>

10 Ibid.

11 Algappa, n 1, pp. 83.

12 Wu Xinbo, n 1, p. 122.

13 Ibid., p. 123.

14 China's Defence White Paper, Internet version at <xinhuawet.com>

15 Kenneth W Allen & Eric A McVadon, *China's Foreign Military Relations* (Washington D.C.: Henry L Stimson Center, 1999), p. IV.

16 Jyotsna Bakshi, "Russia-China Boundary Agreement: Relevance for India", *Strategic Analysis* (IDSA), Vol.XXIV, No.10 (July 2001), pp.

17 Jeffrey Record, "Thinking About China and War", *Aerospace Power Journal* (Winter 2001), Vol. XV, No. 4 available at <Airpower.maxwell.af.mil> Doument created on 6 December 2001.

18 Dmitri Trenin, deputy director, Carnegie Centre Moscow, "The China Factor", an undated study from the archives of IDSA, New Delhi.

19 John W Garver, n.7, p.373.

20 Ibid., p.37.

21 Michael D Swaine & Ashley J. Tellis, *Interpreting China's Grand Strategy: Past, Present and Future* (Santa Monica, Project Air Force, RAND, 2000).

22 David Shambaugh & Richard H. Yang (ed), *China's Military in Transition* (Oxford: Clarendon Press, 1997).

23 Swaine & Tellis, n.21, p. 98.

24 Gerald Segal, *Defending China* (New York: Oxford University Press, 1985), pp. 231-256.

25 Conversations with an American scholar.

26 Zalmay M. Khalilzad et al, *The United States and a Rising China* (Santa Monica: Project Air Force, RAND, 1999), p. 10.

2

Instances and Patterns of China's Use of Force

Tibet, 1950

Having signed the 14 February 1950 Friendship Treaty with the Soviet Union, the Chinese Communist leadership wasted little time in embarking on the sacred mission of national reunification. Although this event took place over fifty years ago, it remains important because it completely transformed the Sino-Indian political equation. The first to feel the brunt of the Chinese Communist Military Power was Tibet. Here it is important to remember that although Mao had proclaimed the founding of the PRC in 1949, the Communist Government's control did not extend to the present frontiers of China. It was some time before the Communist leadership consolidated its rule over all of China. Sichuan, the southwestern province was one of the last to come under central control, and it was through this region that the PLA marched into Tibet. The ruggedness of the mountainous terrain and the immense difficulties that the topography must have presented to the advancing PLA forces can be gauged from the fact that the PLA had to cross thirteen major mountain chains before reaching Tibet. Access to Tibet from the northwest through Xinjiang, although circuitous, was somewhat easier than from the east through Sichuan. The PLAAF began airdropping supplies to them in April. As a result, it established its transportation aviation troops, but possessed only one unit with 12 C-46 and C-47 transports located in Beijing. Eventually, the air force deployed six of these aircraft and four others from southern China to Chengdu to carry out the air drop operations. "This unit soon acquired several Il-12 transports from the Soviet Union and changed its name to the 13th Air Division. From April 1950 to the conclusion of air operations in November 1952, the air force opened up 25 navigation routes across the Tibetan plateau; flew 1,282 sorties; and dropped 51 tons of supplies."[1]

The PRC leadership soon realised that without opening up an all-weather road to Tibet through Xinjiang it would be extremely problematic to rule over this inaccessible region. It also became clear that the PRC's control over Tibet would become untenable without proper logistics and surface communications. Mao and his advisors must have also known that the proposed road would pass through Indian territory and that the PRC would need Pakistan's help to secure its lines of communications to Tibet.

Tibet's absorption, however, was a foregone conclusion. Soon after the declaration of the People's Republic in October 1949, official government statements had made it abundantly clear that Tibet would be '*liberated*'. Military preparation began in early 1950 but took some time, as some portions of Sichuan were still not completely under central control. The laying of roads towards the Tibetan border began only after Garze in Sichuan was captured in April.[2] The PRC began serious military preparations simultaneously with attempts at peaceful unification. Military coercion was thus a critical part of Chinese strategy. In January 1950, the Dalai Lama was ordered to send a delegation to Beijing for talks and to cease attempts to obtain international support. In February, a Tibetan delegation left Lhasa for India (that was the only way it could have travelled to China). By April, when Chinese military preparations were underway, an eight-man Tibetan delegation tried to reach China via Calcutta and Hong Kong but could not as Britain did not grant the delegation a visa to fly to Hong Kong. Britain was in its own way impeding the PRC's takeover of Tibet. By the middle of August, however, the British withdrew their refusal to grant visas to the Tibetan delegation. By the end of the month, the Chinese leadership was getting impatient and demanded that the Tibetan delegation reach Beijing latest by the middle of September. The PRC also let it be known that there was little time for negotiations as "PLA plans were well underway".[3] The Tibetan delegation was already in India awaiting visas to go to China via Hong Kong. Talks with the Chinese charge d' affairs in Delhi and later with the newly arrived Chinese ambassador were unsuccessful. The Chinese government now surmised that the Tibetans were trying their "delaying tactics". The Chinese again warned, this time the Indian government, that the Tibetan delegation must quickly arrive in Beijing for talks even though Chinese forces invaded Tibet on 7 October 1950 without waiting for a concrete Tibetan response. It is interesting to note that the terrain forced the Tibetans to use the India route to proceed to China and whatever the Indian position on the issue the Chinese chose to warn India and not just the Tibetan leadership. On 24 October 1950, the Tibetans said that they would, in fact, go to Beijing but a day later, the Chinese media announced that Tibet had been invaded. It took the PLA just twelve days to capture Qamdo on 19 October 1950.

Several Tibetan delegations representing the Dalai Lama, the pro-China Panchen Lama, and another from a Chinese Tibetan group began talks in Beijing in April 1951 and signed an "Agreement on Measures for the Peaceful Liberation of Tibet" on 23 May. The Chinese representative Zhang Jingwu entered Lhasa on 17 August and the vanguard of the PLA troops from Qamdo reached Lhasa on 19 September 1951.

The liberation of Tibet was, thus, complete. The PRC formed the Tibetan Military Region in February 1952.

It was a relatively minor military operation staged by about 30,000 PLA troops and the casualties were light; about 180 killed with 859 taken prisoner on the Tibetan side and about 2000 on the Chinese side, mostly due to the geographic and weather conditions. From the Indian viewpoint, this small military operation by the PLA was of immense geo-strategic and geopolitical importance. It was not "merely part of the broader anti-Guomindang war of liberation" but the "PLA campaign in Tibet was the first use of force by Communist China beyond its frontiers". Tibet's status as an integral part of China was thus not a 'given' at the time and the PRC leadership was determined to establish its sovereignty over this distant but strategically important region flanked by India on the south.[4] China's control of Tibet immediately brought its military presence on India's northern boarders and in one stroke removed the buffer between the two Asian neighbours.

The Indians found themselves in a difficult situation. According to Garver, there was no difference between the perception of Sardar Patel, the then deputy prime minister and home minister of India and Nehru, the prime minister. Nehru was clear that the PLA's march into Tibet was unstoppable as India was weak and no major power was interested in Tibet's independence. Even so, Nehru was keen to safeguard to the extent possible Tibet's religious and cultural autonomy under Chinese sovereignty.[5]

The most important and urgent lesson of this PLA campaign was that an alternative land route to Tibet had to be found. The difficulties in moving large bodies of troops over the rugged mountainous terrain would always make the PRC's hold over Tibet tenuous. The Chinese soon started preparing to build a highway that eventually linked Tibet from the west through Xinjiang. Although circuitous, this highway guaranteed an all weather access to Tibet. The strategic importance to

China of this highway that eventually passed through Indian territory needs no emphasis.

The 1954 India-China Trade Agreement under which India gave up its special rights in Tibet is a clear indication of India's acceptance of the ground reality. In his meeting with Zhou Enlai in 1956, "Nehru understood that China was also ready to uphold Tibetan autonomy with Dalai Lama as leader but Zhou had warned that 'foreign subversive activities directed against Tibet', would compel China to increase its control there." The activities of the US Central Intelligence Agency (CIA) and the Nationalist-Chinese in providing arms and equipment to Tibetan rebels by over-flying Indian territory in the early 1950s further vitiated the atmosphere. It is possible that the Indians were unaware of these clandestine activities. Nehru's advice to the Dalai Lama to 'peacefully and non-violently' resist Chinese attempts to reduce Tibetan autonomy "was a further demonstration of India's duplicitous policy".[6] In short, Nehru had tried to remove the Chinese military threat by non-violent means but did not succeed. He had hoped that the serious logistical difficulties of maintaining the PLA in Tibet would generally force Beijing to give *de facto* autonomy to this hermit kingdom, but the 1954 rebellion by the Tibetans in Sichuan only reinforced Chinese concerns and determination to further consolidate PRC's control over Tibet.

Another feature of this PLA move into Tibet was that until then India's northern boundary was with Tibet. There was no urgency for its delineation as traditionally it had been a largely peaceful border. With Tibet becoming an integral part of the PRC, there was no guarantee that the Chinese would honour the traditional border. It was this fear that had motivated Sardar Patel to write to Nehru soon after the PLA's foray into Tibet. In hindsight, it seems clear that India a weak and almost pacific power did not appreciate two important facts. First, China was unlikely to brook any outside interference in Tibetan affairs, howsoever noble and peaceful the intentions.

Second, India was in no position to control the actions of the Tibetans in their struggle for independence. The 1959 revolt and the Chinese crackdown following which the Dalai Lama fled to India with tens of thousands of Tibetan refugees forever soured Sino-Indian relations. The major lesson to India was that a weak power could not hope to deal with a large and militarist neighbour if the latter was determined to expand its control; a lesson that India finally learnt only after the ignominious military defeat in 1962.

Korea

The Chinese decision to go to the aid of Communist North Korea was taken at the same time that the Tibetan drama was unfolding. At the time of its entry into the Korean War China, North Korea and the USSR were all part of the larger Communist alliance but even so, it is not clear if China had in fact approved of the North Korean plans to invade South Korea.[7] Writing in 2001, Garver, however, seems certain of Chinese approval of the North Korean plans. He says, "China exercised its leadership of the Asian revolutionary movement in two countries, Korea and Vietnam. In Korea, Mao gave his critical approval to Kim Il Sung's 'war plans', supplied crucial manpower for that offensive and then undertook a very costly three-year war with the United States once US intervention defeated Kim's initial plan".[8] This shows China's readiness to fight a war on behalf of her smaller neighbours. According to Segal, "No war was as important to the PRC as the Korean war". The Korean War was unique in many ways. The Chinese suffered unprecedented casualties. It was the only war in which all three Chinese armed services participated; and the only war in which the PLA troops were forced to retreat. This was also the only prolonged conflict that the PRC fought with a superpower. The war also had considerable influence on the basic elements of Chinese defence and foreign policy".[9] In fact, it further fuelled Chinese xenophobia.

Although military preparations had gone on for some time and large bodies of troops had been moved to the Manchurian border, according to one view, "the Chinese attack in force at the end of November 1950 achieved tactical surprise, with devastating effects on the US forces."[10] The PLA, under Marshal Peng Dehuai, scored a major victory by pushing the US forces back to the 38th parallel. They then halted the offensive, and even freed American Prisoners of War (POWs). These actions lulled the Americans into believing that China was not in fact prepared to fight the US if it decided to advance towards the Yalu. The American aim obviously was to reach the Yalu River the North Korean border with China and end the war with the unification of the Korean Peninsula. The Chinese, on the other hand were determined to ensure the continued existence of their North Korean ally if they could not establish a larger Communist Korea. The first Chinese actions were, thus to deter US interventions and "the PRC's foreign minister Zhou Enlai had, through the Indian ambassador KM Pannikar warned the US of China's intent to intervene to prevent a US threat to China's North-East".[11] President Truman did not take this warning seriously and, in fact, ordered the American Seventh fleet to patrol the Taiwan Strait, which further threatened China and prevented it from completing the process of the liberation of the country by taking Taiwan.

By encouraging Mac Arthur to review his offensive towards the Yalu, the PRC was following the classical Chinese stratagem of 'luring the enemy deep'. This is not to suggest that PLA fought the war entirely under the "people's war" doctrine. The PLA resorted to positional warfare and made full use of its artillery when faced with US superiority in tanks and air power. The PLA resorted to deception by leaving many damaged bridges undisturbed and using temporary bridges under the cover of darkness and submerging them under water during day light. The American reconnaissance failed to detect this. It seems clear that Mao's people's war strategy also incorporated many of Sun Zi's principles such as "attacking enemy's strategy". Sun Zi had said, "All warfare is based on deception; when capable, feign incapacity, when active, inactivity, and pretend inferiority and encourage arrogance and finally, attack when he is unprepared and sally out when he does not expect you."[12]

Having broken contact with the US Army on 7 November, the PLA paused for almost three weeks to give the Americans a chance to pull back but when the US forces resumed their advance on 24 November, the Chinese counter attacked on 26 November. According to Segal, China pursued multiple objectives, which kept changing in response to changing conditions throughout the course of the war. The Chinese objectives were, achieving swift and relatively less costly unification of Korea with minimal US response, support to wider Communist aims but not at the cost of China's reunification and national reconstruction.[13]

In fact, the Chinese did not achieve any of these objectives. US involvement in the Pacific in general and Taiwan in particular became permanent. The war showed that China could take huge risks and heavy casualties and also that China was ready to fight a super power if it appeared to threaten the security of the PRC. China perhaps succeeded in keeping the American threat away from its borders but failed to deter US interventions in the region. The implicit US threat of using nuclear weapons and the PLA's inferiority in weapons and technology spurred the PRC's efforts towards nuclearisation. "China's political objectives changed several times, from hopes for victory, to deterrence, to defence, to victory again and then finally to settling for a stalemate."[14] China, however, demonstrated its resolve to fight a technologically superior enemy. From then on, no country was likely to trifle with China.

Taiwan Strait Crises, 1954-55 and 1958

As noted earlier, Chinese attempts at unification were roundly thwarted by the American backed, and Chiang Kai-shek led Nationalist forces based on Taiwan, an island some 150 km from mainland China's Fujian coast. There were many other small islands in the straits, the control of which was regularly contested by the opposing sides. In a sense, China's use of force against Taiwan had different objectives at different times. Sometimes these were limited to wrest the control of small islets close to the Chinese mainland, e.g. Quemoy and Matsu, whereas on other occasions it was to create a sense of crisis when US-Nationalist action seemed imminent. The two other major objectives were to "probe US intentions and reactions in terms of regional and international instabilities" and to maintain constant pressure on the US and Taiwanese authorities lest they became too comfortable and entrenched on the island.

The two crises in the Taiwan Strait were more political than military although the PRC resorted to the use of limited and well-calibrated military force. During 1950-53, the Beijing remained preoccupied with the Korean War but sporadic incidents continued in the Taiwan Strait. In July 1954, China once again turned its attention to the "anomaly of the unfinished civil war" by periodical military manoeuvres against enemy-held islands and by frequently declaring its intentions to regain Taiwan and the assorted islands.[15]

In February 1954, China, apparently without any real provocation, began more aggressive naval patrolling near the Dachen islands. China's shelling of the offshore island of Quemoy coincided with the first meeting of SEATO, but other than perhaps a signal of China's unhappiness at the US effort to form anti-China/USSR alliances, the atmosphere was generally calm and normal. Tensions rose only when on 2 December 1954, the US signed a mutual defence pact with Taiwan/Republic of China (ROC). This followed a strong Chinese statement on 8 December, warning the US against interfering in China's civil war. China also dropped some leaflets on the islands of Dachen and Ishan asking their population to surrender, but did not actually launch a military operation. On 5 February 1955, the ROC quit the militarily indefensible island of Dachen, which China promptly occupied on 13 February. Since the US had already advised the ROC to vacate the islands, there was little tension but China probed the limits of US commitment by occupying the islands and used the opportunity with alacrity. On 26 February 1955 China seized another offshore island and military engagements continued well into 1955. "The so-called crisis, however, ended as mysteriously as it had begun with desultory military action and continuing political uncertainties, but with China having tested the limits of US resolve and having made its point that Taiwan was still a part of China".[16]

The 1958 crisis also had no clear start or finish and little sense of threat that it might escalate into a real crisis. It was once again an attempt to probe US intentions in its commitment to defend the ROC. Following a political campaign pledging heightened vigilance over Taiwan, and a period interspersed with Khrushchev's visit to China in end July, nothing happened for a while till China suddenly began military operations against Quemoy. On 28 August, the Chinese radio stations broadcast a warning of imminent landings on Quemoy and on 4 September, China declared a twelve-mile territorial waters limit for all its coasts. On the same day, US Secretary of State Dulles warned China that the US would use force if necessary to keep Quemoy from falling to a Chinese blockade. Zhou Enlai cleverly "proposed the resumption of ambassadorial talks in Warsaw with the US". This crafty manoeuvre once again brought events back into China's control by reducing the dangers of escalation, while allowing for continuing military pressure on Quemoy and a useful morale raising campaign at home. Intermittent Chinese firing continued, with hopes of the island surrendering without undue risk. The balance then shifted in Taiwan's favour; the Chinese blockade could be regularly broken and newly acquired US equipment such as the Side-Winder air-to-air missiles took a heavy toll on the Chinese MiG fighters. On 5 October 1958, China proposed a *de facto* ceasefire so long as the US stopped 'convoying Taiwanese ships to Quemoy'. China now shelled the island on alternate days before it finally stopped, ending the crisis.[17]

In this case, China had used military force to convey its displeasure, as well as resolve to liberate Taiwan. It also demonstrated the PRC's propensity to probe US limits while at the same time carefully avoided escalation. Thus, it was a political cat-and-mouse game. According to Segal, "China had used its military for political signalling and stopped when it appeared certain that the Taiwan-US combine will not permit China to occupy Quemoy." Although Mao once again pronounced that "China's millet plus rifles, were superior to American Atom Bomb plus planes", the crisis had once again brought home the fact of Chinese technological inferiority, especially in air power.[18] These crises once again demonstrated the PRC's proclivity for risk-taking and for keeping the Taiwan issue alive lest the world forgot the importance of its unification with PRC.

India

Much has been written about this humiliating episode in India's independent history and only a brief discussion of China's actual use of force should suffice without a detailed examination of the political issues involved in this short but decisive military operation of the PLA. Of all the writings on the Sino-Indian conflict of 1962, Neville Maxwell's *India's China War* seems to have been very popular with both Chinese and Western readers. This should not come as a surprise for Maxwell has very successfully managed to render an accurate and credible account of the war, repeatedly highlighting India's refusal to accept an apparently reasonable Chinese proposal to negotiate the vexed boundary issue. He also gives much credence to the difficulties faced by a parliamentary democracy in resolving issues that get highly emotive when debated between the treasury benches and an alert opposition on the floor of the Parliament. Maxwell, however, is at times more anxious to condemn Nehru's forward policy than to focus on Chinese military and political game. It must be reiterated here, that the forward policy that Nehru put into effect in the end (that is in 1961) was, in fact, nothing more than an attempt to show the Indian flag in areas which were justly claimed by India but where Indian military or political presence was minimal or in places altogether absent. That China should see this as provocative Indian encroachments into Chinese territory is stretching the imagination. The only occasion when the local field commander appears to have been somewhat less discreet was when he decided to draw the boundary at Longju as a straight line from the West to East without bothering about the Chinese claim line.[19]

India, no doubt, erred on two counts. First, his advisors led Nehru to believe that China would never attack and, hence, the "bark but will not bite", statement. Second, and perhaps the most important, was Nehru's inability to see the Chinese anxiety about an unstable Tibet. The PRC leadership was somehow convinced that India was, with the active help of both the superpowers, determined to reverse the *status quo ante* in Tibet. All that India had hoped to achieve was a reasonable degree of autonomy for Tibet and if possible a peaceful border without Chinese military presence. There is no doubt that India did not want war, and perhaps nor did the PRC, but it is equally clear that once the Chinese had determined India's utterly poor military preparedness, they decided to launch the attack anyway to first administer a political and psychological shock and then to teach India a lesson for meddling in the 'internal' affairs of their country. It was also at this stage that, "the superpowers, having emerged distinctively shaken from the Cuban missile crisis, took interest in the Himalayan events". Chinese compellence of India to withdraw from its forward policy was begun ruthlessly on 20 October. "Indian troops withdrew swiftly, but in the three week lull after the PLA offensive, India made it plain that it was bloodied but not bowed"[20]

"China launched a new offensive on 18 November 1962 and this time the panicked Indians fled to the foothills in the East but gave a worthy account of themselves in the West even where they were hopelessly outnumbered. China withdrew as quickly as it had intruded, while Indian pride prevented its leadership from formally accepting the Chinese ceasefire proposals, India, in fact, took care not to provoke the Chinese into another attack."[21]

Some four decades on, it is fashionable to talk of the conflict as resulting essentially from misperceptions of each other's true intent, miscalculation of the military strength and resolve, and poor political signalling by both Mao and Nehru. Though largely correct, it cannot mask the essentially expansionist attitude and approach of the Communist Chinese regime.

Neville Maxwell would have us believe (and hence he was quoted by Zhou Enlai in his talks with Nixon a decade later) that it was Nehru's obduracy and even duplicity that was the main cause of the conflict. While there is ample evidence to suggest that India failed on the diplomatic and military fronts, it does not, however, mean that all blame should be laid at India's doorstep.

China was at best dubious when Zhou did not purposely clarify the issue of Chinese maps depicting large tracts of Indian territory under Chinese control. Nehru had raised the issue early enough, but was gullible in accepting the Chinese disingenuous explanation that their maps were old and needed revision. Maxwell simply wondered why Zhou Enlai did not come clear on the issue at that time. The real reason for

China's duplicity in 1950-51 was perhaps China's need to have a friendly India on its southwestern boarders while consolidating its control over Tibet, and to avoid opening another front while its forces were engaged in North Korea. China was also worried that India's widespread cultural influence and strategic potential in the region might some day translate into genuine strategic power. It is for this reason that Strachey said that "the emergence of a new Chinese Empire depended on the single most important factor of India's success as a developing democracy."[22]

In the early 1960s, China smoothly normalised relations with Burma and Pakistan by resolving the leftover border problems. China showed unusual magnanimity and alacrity in securing the1960 border treaty with Burma. With Pakistan, it signed an agreement under which it received on lease, parts of the Indian state of Jammu & Kashmir (J&K) that Pakistan had forcibly occupied in 1947. China, no doubt, included a clause about the arrangement being finalised only once the dispute over J&K was resolved between India and Pakistan, but its intentions of foreclosing India's future option were clear. Signed as it was at a time when India was licking its military wounds and coming to terms with the monumental humiliation it had suffered at the hands of the Chinese, the treaty in one clear diplomatic stroke gave China direct access to Pakistan, secured its southern front along the newly built Karakoram highway and more importantly assured a seat for Pakistan in any future negotiations with India. China's aim of teaching India a lesson was thus achieved with India's options of helping the Tibetan cause closed for the foreseeable future.

One wonders if China's fears of a resurgent India were not exaggerated. Imagine for a moment a situation in which China had actually resolved the boarder dispute without recourse to arms. China could well have secured the control of Aksai Chin under some, mutually acceptable arrangement such as 'lease' and allowed India a face-saver. India would then have remained a useful friend in South Asia. Given the almost visceral dislike for violence and the military, the Indian leadership, in all probability, would never have built its armed forces and even the Kashmir issue could have been resolved with China's active assistance. What China gained by rubbing India's nose in the dust was only a temporary peace on its borders. The Chinese leadership had perhaps made a strategic miscalculation of monumental dimension. But this is mere speculation.

During the crisis, China's political signalling was not always clear and unambiguous. Zhou Enlai kept repeating China's intentions to find a negotiated settlement without once acknowledging that it could have made a mistake in building the highway through Aksai Chin without first checking if the region was indeed within Tibet. China also did not show any inclination to vacate its illegal occupation but, in fact, blamed India for demanding negotiations over what was clearly not Indian. "What is mine is mine, what is yours is negotiable" appears to have been the Chinese approach. Further, China did not raise the issue at the time of signing the 1954 Trade Agreement. Chinese scholars often blame Nehru's forward policy for the unnecessary escalation to war, but conveniently gloss over the Chinese occupation of areas that were clearly disputed for nearly ten years before India's half-hearted military reaction, as demonstrated by its forward policy.

Writing in a relatively recent study, Allan Whiting has commented on China's strategy of risk management:

"Following the perceived success of PLA deterrence in the Taiwan strait, Beijing moved to coercive diplomacy against increased Indian assertiveness along the disputed border. On 13 September it warned New Delhi, 'He who plays with fire will eventually be consumed by fire' and proposed talks beginning on 15 October. Beijing repeated this deadline three weeks later warning India to 'rein in before the precipice' but without waiting for an Indian response, China opened its attack on 20 October. When Indian resistance collapsed, the PLA waited for three weeks to consolidate and improve its logistics, open new roads to envelope the Indian Army and to see if India would show its readiness for talks. Forty years on, one wonders, how the PRC leadership expected India to meekly accept the Chinese offer for negotiations after the latter had attacked Indian military positions. The Chinese, of course, called it a 'counter attack' provoked by Indian encroachments. Zhou Enlai wrote a private letter to Nehru on 27 October demanding negotiations but when New Delhi remained adamant, the PLA resumed its offensive and swept down to the undisputed border along the foothills. On 21 November, Beijing announced a unilateral ceasefire and restoration of the status quo ante provided that New Delhi made no forward moves for three weeks."

"Its coercive diplomacy having failed to move Indian patrols back or to bring Indian negotiators to the table, Beijing took four steps to minimise the risk of escalation. First, Chinese troops were publicly identified as border guards; second, three weeks separated the first and second offensives. As in Korea, this allowed infiltration of mountain ridges along Indian logistical lines for envelopment of enemy units, while cutting of their line of retreat. This also permitted Beijing to observe whether a Soviet response to the initial attack might strengthen Indian resistance. Third, the PLA exercised strict command and control over fighting at both ends of the boarder. Fourth, China announced a unilateral ceasefire, thereby, forestalling any effective US response to Nehru's appeal for air support. Risk management proved completely successful as did military tactics."[23]

Segal says, "Chinese objectives in the Sino-Indian war were overwhelmingly based on a *perceived* vulnerability to Indian territorial encroachments. Thus, China sought first to deter India, and when that failed, decided to compel an Indian retreat. Unlike in Korea, where China faced a far more impressive military power, Beijing could obtain its wider objectives in 1962 because Indian power was so feeble. It is equally clear that, as in Korea, China decided to adopt a *forward defence* 'at the gates' rather than be content to allow a potential enemy to be 'lured in deep'. Forward defence was specially required not so much because the Indian threat was powerful, but more because China's hold on Tibet was weak. China's political signalling was also not very clear. "China failed to respond to the 19 September Indian note until 3 October and even then it was an ambiguous reply."

"In July 1962, the *People's Daily* had warned about going to 'the brink of the precipice' and the Ministry of Foreign Affairs declared that China 'can by no means sit idle', but neither statements reflected any real increase in tension."[24] This system of conveying warnings through official statements, Foreign Ministry notes and numerous indirect statements in the official news media is confusing but given China's 'One Party' system, it is unlikely to change even in the future, placing an added burden on the opponent to decipher Chinese intentions.

The main objective of the PLA attack was to teach India a lesson about the power differential between the two neighbours and further, to show to the wider developing world that India was not powerful enough to claim its leadership. Lessons were also no doubt intended for other, more important, new spectators–the superpowers. "Recently emerged from the brink of the nuclear abyss in Cuba, the superpowers could be shown that emergent China was a power to be reckoned with."

Even during the three-week lull in fighting, the Chinese continued to harp on their readiness to negotiate but were in fact using the time to beef up their logistics and communications. "The scale of losses was also relatively small. While Indian dead and missing totalled about 3000, the PLA losses are thought to have been much less. China captured

3968 Indian soldiers, but not a single PLA man was captured. The Chinese had superior training, which allowed their superior equipment, especially artillery, to be used with devastating effect. This is a somewhat exaggerated assessment. The PRC had in fact, used hardy hill people and frontier guards from Sichuan and adjoining areas. These troops were used to fighting at higher altitudes and possessed winter clothing, which the Indian soldier did not. The Chinese possessed superior intelligence as their agents often entered Indian territory with the help of the locals and hence were able to anticipate every Indian move. The so-called 'human wave' tactics were used only to capture isolated and crucial positions.[25] Finally, the Chinese held the initiative and made full use of it.

"The only important limit on the ability of the PLA to attack was the difficult logistical problems of the Himalayas. After the 20 October offensive, the three week lull was necessary for China to build new roads and bring up new supplies. Indian forces fled faster than the Chinese advance. The combat was entirely between ground forces. Air power was not used by India for fear of reported Chinese superiority in this department. India also feared that employment of air power might bring attacks on the more vulnerable Indian cities. In fact, once the combat was underway, it was suddenly, and surprisingly clear to India that it was a poor match for PLA fire power."[26] This was the only war in which India did not employ its considerable air power and also the only war which it lost.

Segal adds, "It would be nice for the Chinese military strategists to claim that this favourable state of affairs was due to their intelligent planning and superior strategy. Unfortunately, Chinese success was attributable more to, Indian mistakes and weaknesses than to Chinese strengths". The two superpowers also failed to deter China from attacking India. Commenting on the Chinese domestic scene Segal says, "It is undeniable that China did have some serious internal problems, including a revolt in Tibet and a flood of refuges across the Xinjiang border into Soviet Asia. But these immense and disturbing difficulties occurred well before the 1962 crisis with India. Domestic difficulties did not play an important role in the PRC's decision to attack India. It was their confidence in the PRC's military power and a new determination to subdue the Indians, that encouraged the Chinese decision making. Previous procrastination and uncertain signals may well have had much to do with unrest in China, but the main portion of the Sino-Indian crisis occurred under conditions of growing unity and security in China. The essential flexibility in China's use of military force in foreign policy was vividly illustrated in the Sino-Indian border war. China perceived itself to be under threat from India's border encroachments and, on the basis of recent history of domestic unrest China exaggerated the importance of that threat."[27]

Vietnam 1965-68

Remembering the lessons of Korea and the fear of US encirclement of the PRC, Beijing decided to support Hanoi against US attacks on North Vietnam in 1964 and continued through the following years. Whiting says, "Massive deployment of China's best fighter planes within 300 miles of the border, new airfields to serve North Vietnamese aircraft, interlocking radar grids between the two countries, and joint exercises over a 12-mile space below the border signalled Beijing's commitment to the defence of the Hanoi-Haiphong area." In March regular (US) bombing slowly and systematically moved up toward Hanoi, testing Chinese responses at successive latitudes, but no response occurred. Deterrence had stopped initial attacks on Hanoi and Haiphong, but ultimately failed to halt the bombing of all North Vietnam except for a narrow sanctuary along the border. Even so, Hanoi-Haiphong area remained essentially out-of-bounds for US air attacks.

Hanoi claimed that Beijing had initially agreed to provide pilots and planes to defend it but China's support remained restricted to air-defence operations. Then on 16 July 1965, China reneged on its promise of air support. The same month, however, Zhou Enlai instructed PLA railroad engineering divisions to enter North Vietnam. He said that the US goal of a triple encirclement of China had failed in Korea, remained in Taiwan, but now "they have come back to Vietnam." "Deployment began of 20,000 engineering troops to the northeast coast, completing defences there by October and deploying on to the Red River Delta. On 29 July 1965, President Lyndon Johnson increased US troop strength in Vietnam from 75,000 to 125,000, doubled the monthly draft call, and indicated that further forces would be sent as necessary." Ultimately, 320,000 PLA anti aircraft, railroad engineering, and logistics troops in a systematic rotation served in Vietnam between June 1965 and March 1968, when President Johnson announced a reduction of the bombing and the start of peace talks in Paris. The peak level of 170,000 PLA troops came in 1967. Reportedly 1,500 died, and 4,200 were wounded before their final withdrawal in August 1973. Of these, 280 dead and 1,166 wounded were in anti aircraft units and claimed hundreds of US aircraft downed and damaged. These deployments were purposively visible to US intelligence. In addition to bolstering North Vietnam's defences, they served to deter Washington from a decision to invade the North. "Risk management to limit the chances of escalation marked Chinese behaviour throughout its intervention in North Vietnam."[28] It is clear that in its support to North Vietnam, the PRC not only conveyed its resolve to defend its territory against a possible US attack but restricted its involvement, thereby controlling escalation and in the final analysis leaving Hanoi to fight its own battles with the US. Chinese objectives were modest and were achieved relatively easily because the US also cooperated by not committing ground forces in North

Vietnam. This was also the PRC's first instance of use of force after having developed its own nuclear weapons.

Usuri River Conflict with the Soviet Union, 1969

This was the first case of two nuclear weapon powers fighting a limited conventional conflict. This was also the first ever armed clash between China and its long-time benefactor, the Soviet Union. However, in purely military terms, it was not a major conflict. The Chinese leadership had raised in 1963 the question of "unequal treaties imposed during the weak Qing dynasty", with the Soviet Union. Border talks ended with Khrushchev's fall in 1964. The Soviet Union's 1966 treaty with Mongolia allowing the stationing of a large body of troops in Mongolia and reinforcing its military presence along the Sino-Soviet border (which China matched with its own deployments by 1967) and the Soviet invasion of Czechoslovakia in 1968 are some of the more important causes of the clashes in 1969. In particular, the proclamation of the Brezhnev Doctrine of 'limited sovereignty' seems to have raised fears of a Soviet attack on China. China's hydrogen bomb test of 27 December 1968 was another indication of China's fears and deterrence signals.

Clashes along the Usuri River at Damansky Island called Zhenbao in the Chinese language began in January 1969. "On 2 March the first of two major clashes at Zhenbao occurred when apparently, a Soviet attempt to halt the leading Chinese patrol was ambushed by the Chinese covering patrol. Casualties were high and the Soviets seemed to get the worse of the exchange. On 15 March, the Soviet Union apparently "taught the Chinese a lesson" by orchestrating a second incident at Zhenbao, where superior Soviet manpower and firepower decimated Chinese troops." The initiative for talks came from the Soviets with Soviet Premier Kosygin trying to reduce tensions through a telephonic contact but was rebuffed by the Chinese. "It was only on 24 May that China finally responded to the 29 March Soviet proposal for general talks but without setting a date." The Soviet proposal for collective Asian security at the Moscow conference of Communist Parties aimed at China definitely vitiated the atmosphere and months after the beginning of some moves towards negotiations, major clashes again broke out on 13 August in Xinjiang, apparently at Soviet instigation.

On 3 September 1969, the North Vietnamese leader Ho Chi Minh died and many Communist leaders came to Hanoi. Ho's last testament called for Sino-Soviet reconciliation and on 11 September, Zhou and Kosygin met at Beijing airport. According to Segal, "Kosygin had already reached Soviet Central Asia on his return journey before turning back to Beijing. Apparently the decision to talk was not taken easily in China."

In September, the Soviets warned of nuclear dangers with the Chinese announcing the two nuclear tests of 23 and 29 September. Nuclear sabre rattling thus went on for some time though the tension had eased. This was a classic case of two giants skirmishing but not letting the fight get out of hand because of a realisation on both sides that a full-scale war was not in either country's interest. Talks finally opened on 20 October 1969. The crisis was thus more political than military with few military objectives on ground. China wanted to prove that it would not take the Soviet threat without a robust response and that its deterrence had in fact worked. The Soviet Union on the other hand did not want to allow China the luxury of bleeding it through many small-scale confrontations.

According to Whiting, "The context of the planned ambush strongly suggests, however, that its aim was deterrence against a perceived threat of Soviet intervention in China's Cultural Revolution disarray. As in 1962, the perspective of 'trouble within, trouble without' could associate internal vulnerability with the external threat". China's basic objective in the 1969 crisis was to deter perceived Soviet threats. However, the immediacy of the Soviet threat and the level of crisis was China's own misperceptions and aggressiveness. China's paranoia led it to believe there was a Soviet threat to China in the form of the Brezhnev Doctrine (of limited sovereignty) and that a forward policy of deterrence would disabuse the Soviet Union of the illusion that China could be treated like the East Europeans. Some 34 years after the event, it is safe to assume that the PRC may well again misperceive such imaginary threats and adopt the policy of forward deterrence. At least in financial terms, the PRC is now the stronger of the two with Russia eager to serve China's hi-tech weaponry needs. There is a new and young leadership in Beijing but when it comes to threats, real or imaginary, China may well behave the same way.

Xisha, 1974

China's seizure of the Xisha (Paracels) Islands in January 1974 was in itself a small event that went largely unnoticed by the outside world. As in the Tibet case, China, aware of its superiority in overall power terms, acted when there was little chance of a major power joining the opposite side. China had not merely reasserted its control over its claimed territories in the South China Sea but had shown its readiness to grab opportunity when it came. The disputed islands were then under South Vietnamese control but China knew that neither North Vietnam, which also claimed the island territory, nor the US would intervene if the weak South Vietnamese government capitulated. As brought out before, China had long claimed sovereignty over these islands but did not possess the power to occupy them. Here too, the PRC tried to dislodge the Vietnamese troops by a show of force, but when that strategy failed, quickly used overwhelming force to achieve the objective with minimum fuss and bloodshed. When South Vietnam reinforced its military presence on Nansha Islands further south of Xisha, China protested but did not follow through since it did not then have the necessary amphibious military capabilities to carry out the operation and also to avoid the risk of escalation. The Xisha action was once again proof of PRC's, "long standing desire to compel others to accept its definition of Chinese territory." Chinese success in 1974 was largely due to the weakness of the opponent and the propitiousness of the moment chosen.[29]

"Writing in its Editorial of 24 January 1974, the *Far Eastern Economic Review* (FEER), "Peking also claims the Paracels Western Sands where the fighting is going on, and Spratleies (*sic*) or Nansha (Southern Sands) as the Chinese call them, and Pratas (Middle Sands) and the Macclesfield Bank Group. It would appear to the layman that China's claim to these smacks of imperialism and that counter-claimants could well use against Peking the argument China itself uses in denying the Soviets right to occupy Chinese territory –that the situation is a result of old imperialist days of unequal treaties."

Vietnam, 1979

The decision of the PRC leadership to teach Vietnam a lesson is the more significant for several reasons. This was the first instance of use of force by China in the post-Mao era. This was an aggressive move albeit preceded as before by ample warnings, against an erstwhile Communist ally. The Chinese leadership was peeved with what it saw as Vietnamese adventurism in Kampuchea (now Cambodia) which had a pro-PRC regime. The decision to invade Vietnam was also to tell the Soviets that China would not tolerate any outside interference in its backyard. This, however, proved to be an abject failure of Chinese military and foreign policy. The PRC learnt the bitter lesson about not embarking on even a limited border war when its military was technologically weak and backward. China realised that it was fighting a battle-hardened foe that had successfully defeated the mighty US. "Arguably, the 1979 Sino-Vietnamese war was China's most important foreign policy failure since 1949, and the main reason for this failure was the poor performance of the PLA. In essence, the failure of policy can be attributed to China's incorrect assessment of the link between the political and military dimension of war."[30] This is a striking comment since the PRC has all along been known for effectively mixing the political with the military in the conduct of its foreign policy.

This war has particular relevance for India because it was launched when the Indian foreign minister (AB Vajpayee, later the prime minister) was on an official visit to China with the avowed aim of improving relations that had soured since the late 1950s. China was also conveying a warning to India, a friend of Vietnam, to stay out of south-east Asia.

Although the roots of this conflict can be traced to history, the immediate cause of Chinese anger was the start of the Vietnamese collectivisation scheme that resulted in a large number of ethnic Chinese fleeing to China from Vietnam, and Vietnam's invasion of Kampuchea in December 1978. China was also opposing Soviet hegemonic designs in the region. Vietnam had signed a friendship treaty with the Soviet Union on 03 November 1978 and the "PRC's August 1978 treaty with Japan had a clause to oppose Soviet hegemonism"[31] The Vietnamese military drive against Kampuchea had begun on 25 December; Phnom Penh fell on 7 January 1979. The Chinese made general preparations for war by moving their forces to the Vietnamese border and Deng himself went to Washington and Tokyo to assess their likely reaction. After his visit to Washington, "Deng apparently felt that he had heard nothing to dissuade him, from his purpose." According to Segal, "There was once again no consistent pattern of escalating Chinese warnings to Vietnam that China would strike, for even the final warning from the Ministry of Foreign Affairs, on 16 February, was no different from earlier statements."

On 17 February 1979, the Chinese PLA forces struck along a broad front across Vietnam's northern borders. Initially their progress was slow but was reinvigorated on 23 February and although slow and costly in Chinese lives, the PLA finally began a major assault on Lang Son, a provincial capital, on 27 February capturing it on 5 March. China thereafter immediately announced its intention to withdraw and by 16 March all Chinese forces were pulled out. Hanoi agreed to hold talks and China agreed to send its delegation to Hanoi by early April.

Chinese military preparations included a major programme of upgrading airfields, logistical routes, and medical facilities in Guangxi Province opposite the Vietnam border. This went for over a month and was almost certainly detected by Hanoi. However, public threats and PLA deployments failed to deter Vietnam from invading Cambodia on 25 December. Phnom Penh was captured in twelve days and China's client state collapsed in two weeks.[32]

Deng Xiaoping's determination to "teach Vietnam a lesson" left China's military goal undefined. Beijing nonetheless addressed risk management much as in the 1962 border war with India. The PLA withdrew after three weeks, destroying five provincial capitals but stopping at Lang Son, a key junction en route to Hanoi. No air power was used. In fact, the PLA moved its supplies and even aircraft spares by road and river craft. The Chinese ambassador remained in Hanoi throughout the invasion. On 26 February Deng told a Japanese journalist that the attack showed China had "no fear of the Soviet Union."[33] What is important from the view point of this study is that the Chinese were ready to launch an attack simply because their smaller neighbour was acting rather independently and even when it had recently signed a treaty with a superpower (the Soviet Union). This clearly shows the PRC's proclivity to use force even when the extant strategic situation is not plainly in its favour. Its decision to withdraw its forces immediately after it had posted a military success, in this case against the provincial capital of Lang Son, once again shows its general reluctance to get involved in a long drawn out war on its border. Unlike in the 1962 conflict against India, however, this time the decision to pull out the PLA was more because of its dismal showing on the battlefield. The PRC, it is evident, learnt a bitter lesson and Deng's efforts to sincerely and seriously address the deficiencies of the PLA began as a direct result of this military failure under his overall command.

Nansha 1988

China refers to the Spratly Islands as the Nansha Islands, and claims all of the islands and most of the South China Sea for historical reasons. These claims are not marked by coordinates or otherwise clearly defined. The Chinese claims are based on a number of historical events, including the naval expeditions to the Spratly Islands by the Han Dynasty in 110 AD and the Ming Dynasty from 1403-1433 AD. Chinese fishermen and merchants have worked in the region over time, and China is using archaeological evidence to bolster its claims of sovereignty.[34] China's seizure of some of these islands in 1988 did not arouse much interest but the 1995 Mischief Reef incidents in which the Chinese and Philippines troops confronted each other raised some alarm. ASEAN and China have at their October 2002 Bali summit signed a non-aggression treaty paving the way for a peaceful resolution of the territorial disputes in the South China Sea. What is, however, significant is that a mere agreement on "code of conduct" that the ASEAN powers had successfully extracted from China in 1999 was obviously inadequate to allay their fears. It is quite possible that in the future, these disputes would erupt again when China decides to exploit the mineral resources in and around these disputed territories.

Taiwan Strait Exercises, 1995-96

This is the most recent demonstration of Beijing's resolve to dissuade any outside power including the US from helping Taiwan inch closer towards independence. This time the PRC chose to intimidate Taiwan and its backers by staging a spectacular missile firing exercise that continued with other military exercises for almost a year. The absence of actual combat by the PLA during the military exercises in the Taiwan Strait in 1995 and again in 1996 differentiates this episode from the previous ones. It, however, illustrates the recurring pattern of deterrence and the Chinese penchant to seize the initiative whenever confronting the United States. In May 1995, Lee Teng-hui, the president of the Republic of China, contrary to Washington's assurances to Beijing, received a visa to Cornel University, his *alma mater*. In June, his speech there emphasised the international status of Taiwan, triggering a wave of rhetorical attacks from Beijing, which were followed by missile exercises north and south of Taiwan.

From 21 July to 28 July, 1995, after public warning, the PLA fired its missiles 80 miles northeast off Taiwan in a 10-nautical mile circular area near the air and sea lanes between Japan and Taiwan. China's objective was to warn Washington against further support for Lee and to deter Lee from continuing his perceived moves toward Taiwan's independence.

The PRC did not stop at that but 'for the sixty-eighth PLA anniversary on 1 August, Beijing's defence minister declared that China "will not sit idle if foreign forces interfere in China's reunification and get involved in Taiwan independence. By August, upward of 400 articles in the Chinese press attacked Lee by name."

"In the fall regular semi-annual PLA exercises took on an expanded format as CCP Chairman Jiang Zemin watched an amphibious landing on a simulated 'enemy shore' in the Yellow Sea, involving aerial combat and missile bombardment of shore- defences while submarine and antisubmarine units blocked the operations from 'enemy' ship attack." In mid-November ten-day long exercises once again simulated an invasion of Dongshan Island near southern Fujian in reportedly the largest combination of ground, naval, and air units in the PLA's history.

Some 160,000 troops participated with 200 landing craft and 100 other ships in the exercises, the political target of which was Taiwan, where the Legislative Yuan election was scheduled for 2 December. Two days before that event, a member of Beijing's Taiwan Research Council declared that the exercises were a "most serious warning" to alert those who thought they could "break Taiwan away from China through so-called 'democratic procedures' with the support of foreign forces"."

"After a month long-deployment that included elements from all three PLA fleets, an estimated 300 planes, and 150,000 troops, Beijing scheduled missile exercises for 8-15 March during Taiwan's three week campaign period. The relatively small target areas were 32 miles from the southwest coast with a major port, Kaohsiung, and naval base Tsuoying, and 22 miles from the northeast coast with a major port, Keelung and naval base, Su'ao."

On 15 March, the NCNA (New China News Agency) announced a third set of exercises for 18-25 March, ending two days after Taiwan's election and covering 6,000 square kilometres around Haitan Island off the Fujian coast. There, "the amphibious boats, helicopters, and parachute forces would combine with land, air, and naval forces to practice seizure of the islands."

"Risk management began in mid-1995 with the explicit designation of scheduled missile firings and the advance announcement of exercises in the fall. In addition, *none of the missiles fired in Taiwan's vicinity in either year were armed.* PLAAF planes flew across the mid strait line in 1996, contrary to the practice by both sides, but they did not come close to Taiwan. Finally, in contrast with all of the previous cases, Beijing took *several steps to inform Washington that no attack on Taiwan was planned.* First, in February 1996, the US embassy in Beijing was informed about this decision. Second, later that month National Security Adviser Anthony Lake proposed that his counterpart come to the United States for discussions; Beijing agreed. Vice Foreign Minister Liu Huaqi met in Washington with Lake, Secretary of Defence William Perry, and Secretary of State Warren Christopher just as the first missiles were to be fired. Frank exchanges that night and the next day established the boundaries of anticipated movement by both sides."

"Unlike previous PLA actions that directly or indirectly involved the Unites States, authoritative high level communication was undertaken to minimize misunderstanding between Washington and Beijing. As an aftermath, in 1997 and 1998 President Bill Clinton and President Jiang Zemin exchanged summit visits."[35] It seems that although Beijing had resolved to prevent a possible "unilateral declaration of independence" by an overly enthusiastic leader in Taiwan, the PRC wanted to avoid armed conflict, if possible. The elaborate arrangements made through direct contacts with high-level officials in the US show this 'missile drama' as a stage-managed show in the classic Chinese tradition. What with the US alerting its Pacific fleets such demonstrations have the potential for unintended escalation but the PRC does not apparently mind taking such risks.

Assessment

Having briefly examined the history of China's use of force, it would be simplistic to conclude that in the future too China would repeat its past behaviour. "In October 1975, Deng Xiaoping told a visiting, American foreign affairs delegation that China could 'wait one hundred years' but would never give up Taiwan."[36] Today, a quarter century later, it is clear that the PRC would never allow Taiwan to become independent, whatever the cost. However, the PRC has much more to lose today than in the past. China's trade and economy is now so intimately enmeshed with the world that unless its core interests were directly endangered, it might not resort to use of force. Military coercion and compellence would, however, remain the most favoured tools of diplomacy. Whether or not it resorts to use of force, would depend on the circumstances. If the PRC leadership were to conclude that passivity or indecision might result in loss of territory, it is very likely that it would use force even when weak. "Thus, the perception of Taiwanese behaviour and US intentions can prompt Beijing to feel being 'backed into a corner'.[37] Similarly, Beijing leadership might resort to force if it perceived a threat to its territorial integrity from dissidents in Tibet or Xinjiang. In all probability, the PRC would strive to maintain a peaceful environment in its neighbourhood while it is engaged in the task of building its national power. *Any attempt to threaten China's claimed territories in the South China Sea, a sudden discovery of large quantities of hydrocarbons or a window of opportunity to establish its control could well prompt the PRC leadership to seize the initiative.*

In the past, internal threats and instabilities have also played a role in China's decision to use force. Whiting states, "In 1962 catastrophic famine after the collapse of the Great Leap Forward triggered alarm and exaggerated threat perceptions with respect to the United States and Taiwan, the Soviet Union and Xinjiang, and India, with the disputed border. In 1969, the political disarray from the Cultural Revolution coupled with a steady Soviet build up along the border sparked the Usuri ambush. A more recent worrisome precedent is 1989, when anger over corruption and inflation underlay widespread demonstrations far beyond the confines of Beijing and Shanghai, where students had begun their marches. Thus, if the 'trouble within, trouble without' syndrome arises from a coincidence of perceived internal and external pressures, the resulting worst-case analysis can impart a greater sense of urgency on winning Taiwan than otherwise would exist."[38] Although here Whiting has focussed on the Taiwan problem, similar behaviour could be expected if trouble erupts in Tibet, Xinjiang or on China's land borders.

A Russian scholar believes that China's behaviour is not aggressive but at the same time, it is "completely capable of over-riding others will with its sheer weight. China will try to prevail without actually engaging in open conflict".[39] It is unlikely that Beijing will resort to traditional military expansion to achieve its foreign policy goals. China is essentially a proactive power but it is not necessarily reckless. The new 'fourth generation' leadership is sophisticated, well-informed yet street smart. When it comes to fighting any threats to the country's hallowed aims of national reunification, Hu Jintao and his team are unlikely to remain passive. They cannot afford to give the impression that they were less enthusiastic about the newfound sense of Chinese nationalism as their regime survival may well depend on it. The recent signing of the non-aggression treaty with ASEAN member countries and the acceptance of Sikkim as an integral part of India are perhaps designed to allay the fears of its smaller neighbours. China's burgeoning economic and military power will, however, continue to pose a challenge to China's neighbours. If it can establish control, coax others to do its bidding and have a generally favourable regional climate, there is little chance of the PRC resorting to force but that does not preclude it either.

China may try its utmost to avoid conflict but it can still not stop an adversary from taking preemptive action to probe China's intentions and limits of patience. In short, what if the adversary actually employs 'China's tactics' of intimidation? In such a scenario, it is quite possible that China would be forced to back off and that would expose its inherent military weakness to the whole world. The PRC always adopts a strategy that allows it to pull back from the brink of a crisis, as in the post-Belgrade embassy strike or the *EP3* (US Spy-plane) incidents. This may not always work and may throw up some unexpected challenges to the PRC if the future adversary shows adequate gumption.

Notes

1 Kenneth W. Allen, Glenn Krumel, Jonathan D. Pollack, *China's Air Force Enters the 21st Century* (Santa Monica: RAND, 1995), p.40.

2 Gerald Segal, *Defending China* (New York: Oxford University Press, 1985), p. 80-91. This information is largely based on Segal and, hence, frequent quotes/references are avoided.

3 Ibid, p. 81.

4 Ibid, p. 81.

5 John W. Garver, *Protracted Contest: Sino-Indian Rivalry in the Twentieth Century* (Seattle: University of Washington Press, 2001), pp. 28 & 54.

6 Ibid, pp. 28 & 54.

7 Segal, n. 2, p. 93.

8 Garver, n.5, p. 116.

9 Segal, n. 2, p. 92.

10 Burles & Shulsky, *Patterns of China's Use of Force* (Santa Monica:, RAND 1999), p. 6.

11 Segal, n. 5, p. 92.

12 Burles & Shulsky, n.10, p. 8.

13 Segal, n.2, p. 95.

14 Ibid, p. 110.

15 Ibid, p. 115.

16 Ibid, p. 116. This account is based largely on Segal.

17 Ibid, p. 117. This information is based largely on Segal.

18 Ibid, pp. 118-119

19 Neville Maxwell, *India's China War* (Bombay: Jaico Publishing House, 1970), pp. 107, 131 and 267. In fact, Lonju is cited more than once as an example of Indian obduracy.

20 Ibid, p.142.

21 Ibid. p. 143-144.

22 John Strachey, *The End of Empire* (London: Victor Gollancz Ltd, 1959), p. 305

23 Allen Whiting, "China's Use of Force, 1950-96, and Taiwan", *International Security*, Vol. 26, No. 2 (Fall 2001), pp. 103-131.

24 Segal, n. 2, p. 144.

25 Ibid, p.146.

26 Ibid, p. 147.

27 This paragraph is from Segal, pp. 146-150.

28 Whiting, n. 23.

29 Segal, n. 2, p. 199.

30 Ibid, p. 211.

31 Ibid. This account is based on Segal, pp. 212-213

32 Whiting, n. 23.

33 Ibid., p. 120

34 "South China Sea Region," United States Energy Information Administration, Country Analysis Briefs, *August 1998* and Greg Austin, *China's Ocean Frontier: International Law, Military Force, and National Development* (Allen & Unwin, 1998) as quoted in http://www.middlebury.edu/SouthChinaSea

35 Whiting, n. 23. This and subsequent paragraphs on the 1995-96 Taiwan Strait crisis are based on Whiting, pp. 120-124.

36 Ibid., p. 127.

37 Bates Gill, "Chinese Military Power", US House of Representatives Armed Services Committee, 19 July, 2000, Testimony, available at <http://www.brook.edu/views/testimony/gill/20000719.htm>

38 Whiting, n. 23, p. 128.

39 Dmitri V. Trenin, "The China Factor", from the IDSA, New Delhi, archives.

3

The Evolution of the PLA Doctrine

Defence or military doctrine is an important indicator of the intentions of a country's armed forces because it tells others how the country intends to use its military capabilities to fight a war. As we have noted earlier, in China it is not easy to make clear distinctions between the three major organs of the state, the central government, PLA and the CPC. The PRC leadership comprising its high-level political functionaries have since its founding worn 'two hats', simultaneously holding two appointments, e.g. state president and chairman of the Central Military Commission (CMC). The PLA and CPC have always enjoyed a symbiotic relationship, with the former pledging its loyalty to the latter and the latter declaring that the Party would always control the gun. Chairman Mao's dominant and long lasting influence over the PLA should, therefore, not come as a surprise. Being the supreme leader of a victorious army, Mao towered over the PLA's strategic thinking throughout his life. Being the founder leader of the Party and the government, he not only dictated national political and military strategy but was also often seen as playing a decisive role in the actual employment of the forces in the field. During the Korean War, the decision to halt the advance in early November 1950 and the release of American prisoners before the launch of the next offensive, and similar decisions in the 1962 Sino-Indian conflict were made on the express approval of the central leadership at Beijing indicating the tight grip of the then Party leadership on the actions and decisions in the field. It was, thus, natural and logical that Mao's People's War doctrine would play a dominant role in China's strategic and defence affairs.

China's closed system and the 'one-size-fit-all' type of defence doctrine of the 1950s and 1960s did not lend itself to clear interpretation because it was often confusing at the best of times. The People's War doctrine not only indicated 'which' wars the Chinese might fight but also that, whatever the odds and howsoever powerful the adversary, the PLA would ultimately vanquish the enemy. In its post civil war history, however, the PLA has hardly ever employed the People's War doctrine.

Four major influences are discernible in the evolution of the PRC's military doctrine, which has gradually evolved over the past fifty years: ideology, huge strategic real estate, lack of economic resources and non-availability of modern technology. This coupled with the absence a solid industrial and technological base exacerbated the situation. The PRC's defence industry which was fashioned on Soviet lines certainly produced large numbers of aircraft, tanks, guns and ships but did not lend itself to easy modernisation and, hence, the self-reliance slogan remained largely unfulfilled. The process of change has no doubt received added impetus with the passing of the revolutionary era leadership (Mao) in 1976, the advent of the pragmatic and visionary Deng Xiaoping and the slow but increasingly spectacular successes of the "four modernisations" programme. Later, the PLA leadership began to review its doctrine once again when modern technologies started to become available with China's increasing economic power and with the gradual cooling off of the international security situation in the late 1980s. The doctrine underwent three major changes: first, in the late 1970s from, "People's War' to 'People's War Under Modern Conditions;" second, in the mid to late

1980s, to "Local Border War Under High Tech Conditions;" and, finally, in the late 1990s, an incipient change towards the use of RMA technologies appears to be gradually emerging.

People's War

Any discussion of the PLA's doctrine must begin with the examination of the famous and almost commonplace term, "people's war", that Mao enshrined in the Chinese political and military lexicon. It must, however, be noted that the concept was not merely a Communist revolutionary slogan but a stark demonstration of necessity. It was after all, the people under Mao that had waged a victorious war against the forces of reaction in the country and, hence, Mao could not have ignored their role in the success of the revolution. What is more important, however, is the continued reliance on the power of the people even after the founding of the People's Republic. This, as John Lewis has said was, "making a virtue of necessity",[1] since without the wholehearted support of the people neither the regime nor national security could have been assured. Communist ideology also played a large part in always highlighting the people's contribution to state and nation building. The term 'People's War', itself needs a deeper look because it has connotations far beyond its use and meaning, as a concept for war fighting. In the post-World War II period, most other countries had switched over to the use of the term 'defence' rather than 'war'. China's or more correctly Mao's insistence to continue to use the revolutionary term shows his fundamental belief in people's power and axiomatically in 'war' as a policy option. In this sense, his thinking was not unlike that of Clauswitz. Mao, in fact, often said that "politics was war without bloodshed and war, politics with bloodshed" indicating thereby his strong belief in the indivisibility of politics and war. It is for the salience that war has always enjoyed in the Chinese scheme of things, that the study of PLA doctrine becomes even more important.[2]

Another aspect that needs to be remembered is that even this hallowed doctrine has been used flexibly in practice right from the very first use of military force by the PRC. It is not as if Mao did not understand or appreciate the vital role of technology in fighting a modern war, but it was because the newly emerging People's Republic had neither the access nor the finances to obtain technology, that it became natural and necessary for Mao to extol the virtues of this concept at every available opportunity. Such demonstration of strong faith of the supreme leader in the power and capability of his people undoubtedly worked as a morale booster to the then impoverished nation and more particularly for an army trying to defend the country with antiquated equipment.

As noted earlier, the PRC entered the Korean War without worrying too much about the actual state of the PLA's equipment, but using the skill and power of its troops, successfully stopped the enemy. The PLA used surprise and deceptions, the two most important tools of "people's war", to great advantage. When necessary, the PLA resorted to conventional positional warfare. With the Soviets rapidly providing the PLA with modern arms, equipment, and aircraft, including the legendary MiG-15 jet fighters, the Chinese showed a reasonably high capacity to use the new equipment and technology that went with it. Although, the PLA suffered heavy casualties, it achieved the primary objective of deterring an American attack on the Chinese mainland. However, it failed to reduce the US commitment in Korea. By its readiness to enter a major conflict against a superpower when the newly proclaimed People's Republic had not as yet stabilised, its leadership gave a strong and clear warning to the world at large and the forces of imperialism in particular that China was not a power to be trifled with; not a mean feat for the PLA and its doctrine.

Mao used the 'people's war' doctrine in a variety of ways such as when dealing with the threat of nuclear blackmail, or when the Soviets withdrew their specialists to send a strong signal that China was not ready to accept its support and assistance on Soviet terms and may well do without sophisticated weapons if it came to that. He, therefore, frequently highlighted the Chinese capacity to fight the superior with inferior, the importance of men over weapons, and downplayed the importance of technology. Debates about the changes in the doctrine to "people's war under modern conditions" actually began in the Fifties but the term was used to exhort the Chinese scientists and engineers to show greater enthusiasm in building a self-reliant yet modern army. In effect, "people's war" doctrine was inextricably entwined in the increasingly strident debate on 'self-reliance'. Since the PRC could not spare the funds for military modernisation, it was essential to repeat the leadership's faith in the self-reliance slogan while simultaneously telling the nation and the world that China was quite capable of looking after its defence on its own. This helped the Chinese leaders to downplay or even denigrate Soviet military and technological assistance when it suited them.

The decision to build its own nuclear weapons was also driven by this major shortcoming (lack of technology). By 1955, Mao knew that the threat of nuclear blackmail could be met only if China possessed nuclear weapons but it would take time to transform that dream into reality. In the meanwhile, China could not appear defenceless and, hence, had to rely on, and repeatedly declare its faith in, the "people's war" doctrine. The PRC not only succeeded in exploding its atomic bomb on 16 October 1964 but within a short span of three years, had also tested the hydrogen bomb and developed a missile force. The PRC had achieved the rudimentary capability to thwart the threat of nuclear blackmail. China could now take its time to develop its conventional defence technology and continue to insist on self- reliance.

"By the early 1980s, the PRC had developed and deployed a small but usable nuclear force."[3] By about the same time, the

Soviets were involved in Afghanistan. This also reduced the Soviet threat to China. China's relations with the US had been developing since the Nixon visit in February 1972 and the general external environment had become relatively benign. Even before the end of the Cold War, Sino-Soviet relations had begun to improve under Gorbachev's leadership. It was, therefore, possible to review the PLA's role and military doctrine in a relatively peaceful atmosphere, which Deng Xiaoping decided to use to push his economic reforms. PLA modernisation was, thus, given a low priority and its funding reduced.

China's geography and cultural tradition also played a major role in the promotion of the 'people's war' doctrine. China's vast geographical expanse and a largely self- sufficient rural economy conferred on it many advantages. First, its strategic depth meant that China could never really be occupied by a conventional armed attack. Second, the PLA could yield ground, lure the enemy deep, and then fight a war of attrition. Mao's famous claim that due to the large population, the Chinese people could not be subdued even if an atomic bomb was dropped on China must be seen in this context. In short, Mao cleverly used the strength of his numerically superior but technologically inferior people's army to deter external threats.

As we have noted earlier, Chinese society voluntarily accepted the central rule, be it of an emperor or a ruling party, only so long as the ruler was perceived as having "heaven's mandate". Hence, it was and even today is, politically expedient to prefix the term 'people' before every national endeavour or activity. It is noteworthy that the PRC did not really use the people's war doctrine in Tibet, Korea, the Taiwan Strait crises, India, and twice in Vietnam (1964-65 and 1979) because it was either on the offensive or the operation was in support of another army fighting a war against imperialism, as in Vietnam. To be sure, the PRC did not always achieve all its political and military objectives in these conflicts but it was only after the disastrous 1979 War against Vietnam that Deng Xiaoping finally, if belatedly, realised the deficiencies of the people's army and its doctrine. It was no longer possible to wage even a limited conventional war without technology. Deng had, in the meantime, already declared that the prospect of "an early, major, nuclear, or world war" had receded. This was tantamount to simultaneously announcing to his people that economic reconstruction was of the highest importance now that the threat of a major war no longer loomed large on China's strategic horizon. Deng, however, said that the possibility of a local border war could not be ruled out. These statements were apparently aimed at his army and China's neighbours. The army needed to improve its efficiency by cutting its flab and the neighbours needed to note that China was still capable of a robust response should they indulge in any adventurism. Implicit in his statement was an assurance to his neighbours that China would not resort to military force unless compelled to do so by their actions.

People's War under Modern Conditions

In the 1980s, the PRC military leadership began a review of the methods and means to win a limited war. At first, the self-reliance debate appeared to go on interminably since there was an increasingly wide gap between resources, capabilities, and ambitions. In the early 1980s, the CMC in its exuberance wanted the PLA to do everything without really ensuring that it had the means to carry out those tasks. For example, it declared that "future wartime tasks for the air force would be to defend strategic points and provide air cover for strategic deployment of mass troops; maintain air domination in the main theatres of operations in support of the army and the navy; launch surprise attacks on high-value targets of the enemy; participate in nuclear counterattack; and conduct strategic aerial reconnaissance". The CMC also expected the air force to "prepare defenses against air raids and support the other services opposing a ground invasion or launching counter-offensives".[4]

It is easy to see how unrealistic these demands were when all that the PLAAF possessed at the time was an assortment of obsolescent bombers and fighters with limited range, and outdated avionics and armament. Deng was quite conscious of the importance of air power for he said,"The army and navy both need air cover. Otherwise the enemy air force will run rampant". He, however, did not declare that he was aware of the PLAAF's deficiencies and that it was in no position to actually carry out these multifarious tasks.[5]

The PLA and the PLAN were in a similar state of inadequate operational preparedness. The PLA's artillery and armour were antiquated and the long neglect of the PLAN had left it barely capable of meeting coastal defence needs. But the PRC had to search for an appropriate doctrine before the process of reinvigorating and modernising the PLA began in right earnest. What with the plans for indigenising defence and downsizing the PLA, it was clear that it could not be everywhere at the same time. The task of improving its fighting efficiency so that the PLA could face any contemporary armed forces was not simple. The changes in the PLA doctrine must, therefore, be seen in this light. It was evident that there would remain gaps between the capabilities and missions but a beginning had to be made.

Writing in 1997, Nan Li felt that "Chinese military planners like to claim that in *overall* terms, the PLA is technologically inferior if compared with its potential adversaries. Such inferiority will continue in the foreseeable future, even though the PLA has made some progress in modernising its weaponry in recent years."[6] While generally true, the statement needs to be taken with some caution because, as we shall see in the succeeding sections, the process of modernisation has come a long way since the mid 1990s. The eight-character principle of "active defence and luring the enemy deep" had undergone a change by the 1980s. "In the autumn of 1980, the military commission resurrected the 1956 concept of having front-line troops hold off the enemy from prepared fortifications until the nation could be put on a wartime footing...abandoning Mao's idea of luring the enemy deep. This new guideline placed positional warfare at the forefront of Chinese strategy." Deng had by 1985, discounted the chance of a major war but the threat of a local border war had to be met. "PLA analysts worried about the optimal approaches to winning local wars between China and its neighbours. They deemed Vietnam and India to be their most probable future adversaries. A new 12 character phrase captured the approved policy for regional conflicts: adopt every diplomatic means to check the outbreak of war; ensure victory if war is inevitable; and end the war on favourable terms after the predetermined goals have been achieved."[7] Before discussing each of the changes cited above in detail, it must be noted that the 1991 Gulf War had an enormous effect on the Chinese military thinking although the PLA top brass initially dismissed the Coalition successes as being inapplicable to their armed forces. Lewis cites an interview in which, "President Yang Shangkun appeared to belittle the relevance of American military might to his country: The model [of the Gulf War] is not universal. It cannot, at least, be applied in a country like China, which has a lot of mountains, forests, valleys and rivers. Another characteristic of this war is that the multinational forces faced a very weak enemy."[8] There is certainly some merit in such assertions but the PRC leadership had also realised that when faced with such a situation where the enemy possessed overwhelming technological superiority and precision firepower, China would suffer almost total destruction in the first wave of attacks and would be left with little to retaliate with. This realisation is, in fact, the basic reason for the adoption of the doctrine that includes pre-emptive attack while officially adhering to the "active defence" slogan.

The following is a good overview of how the PLA's doctrine was changing.

(a) From luring deep to fighting forward.

(b) From a war annihilation to a campaign against key points.

(c) From a war of attrition to a decisive campaign with a decisive first battle.

(d) From waiting for the first blow to deterring the first blow by force.

(e) From a defensive campaign to an "offensive defence' campaign.

(f) From "advance and retreat boldly" to checking the initial enemy advance.

(g) From a "front army campaign" to a "war zone" campaign.

(h) From the principle of mass to the principle of concentration of firepower.

(j) From the four single service campaign to joint campaigns.

These are the attributes of a proactive doctrine much more forward leaning than in the past.[9]

From Luring Deep to Fighting Forward.

This was a guerrilla warfare tactic developed and adopted during the civil war. This, as the term suggests, meant yielding ground, and allowing the enemy to advance, and then engaging him in a battle of attrition. Mao's formulation based on Sun Zi: "When the enemy advances, retreat, when he tires, counterattack; when he retreats, harass; and use overwhelming force in attack", etc. is also deeply founded in guerrilla warfare, deception and surprise.

Interestingly, despite the overarching influence of the "people's war" doctrine, the PRC or more correctly the PLA never yielded ground, at least while in own territory, but gave battle at the 'gates' in all of the nine or ten conflicts that it engaged. In fact, the PLA fought aggressively and even made deep forays into enemy territory when the situation so demanded, as in the case of India and Vietnam, but pulled back when the objective of "teaching a lesson" had been achieved or when logistics so dictated.

From a War of Annihilation to a Campaign Against Key Points.

The PLA, in its earlier experience, relied mostly on infantry warfare since it barely had any artillery support. It had to, therefore, ensure the total annihilation of the enemy, either the Japanese, the Nationalists or other warlords or bandits, in order to secure total victory. In a limited war, however, the objectives would be limited to forcing the enemy to negotiate and hence launching a campaign against "key points", e g where the enemy was weak or where the target was of high strategic value would suffice. There would be no need to launch an offensive on a broad front and fritter away the precious resources. Once the battle for "key points" was won decisively, the PRC could begin to increase political pressure on the enemy to come to the negotiating table.

From a War of Attrition to a Decisive Campaign with a Decisive First Battle.

This is similar to the previous tenet as the rapidly downsizing PLA could not now afford to incur losses in manpower and equipment and further lose territory. It can no longer field large bodies of troops as in the Korean War. It has to, therefore, be ready to take the initiative at the very first opportunity without waiting for the enemy to make is first move.

From Waiting for the First Blow to Deterring the First Blow by Force.

This is a noteworthy departure from its putative "defensive" approach and actually means a strong 'pre-emptive' strike when the enemy's attack appears imminent. With improvements in

PLA's offensive capabilities, especially its armour, artillery, naval power and, most importantly, in its combat air power, the PLA can now plan to mount a joint offensive. In fact, in a 1997 conference of senior military commanders the then president of the Academy of Military Science had said, "The very assembly and deployment of coalition forces constituted the 'first firing' and justified pre-emptive military action. "Such pre-emptive action may postpone or even deter the outbreak of war".[10] Given the prevalence of this thinking, it is probable that in any future war, the PLA would not wait for the enemy to strike first. This thinking is clearly based on the lessons of the 1991 Gulf War, and the other three wars namely, the 1999 Kosovo Conflict, 2001 Afghanistan War and the recently concluded Iraq War which would have only reinforced this belief. The PLA is now vigorously building its capabilities for such a pre-emptive attack.

From a Defensive Campaign to an "Offensive Defence' Campaign.

This once again shows that although the PLA has all along stressed "defence" or now "active defence", its true doctrine has always been offensive. As we have seen in an earlier section, the PLA has used force when necessary and not necessarily in response to an enemy attack. This change should, therefore, come as no surprise to students of China's military modernisation. Even when its modernisation is complete, the PLA would in all probability, have only limited resources and staying power and hence may not carry the pre-emptive very deep into enemy territory except when the offensive was opened with an air strike against the enemy's strategic targets or what are often referred to as the enemy's "centres of gravity". In a classical ground offensive too, the PRC would not like to risk getting bogged down deep inside enemy territory away from its logistics support. In the case of a conflict over territories claimed by the PRC in the South China Sea, however, the PLA would press home its attacks to ensure full and complete control. In a possible future conflict over Taiwan, it would be the extent of US intervention that would determine the Chinese response; but here too, the PLA would try its utmost to deter such an intervention or at least limit it if possible.

From "Advance and Retreat Boldly" to Checking the Initial Enemy Advance

This change is once again a direct result of the PLA's understanding of modern warfare that is dominated by air power. The PRC/PLA leadership know full well that absorbing a 'first strike' by a technologically superior adversary possessing a modern air force would in all probability so weaken the PLA that it would be nearly impossible for it to retaliate decisively. The initiative would, thus, pass to the adversary and, hence, the stress on checking the initial enemy advance.

From a "Front Army Campaign" to a "War Zone" Campaign

This change in thinking is perhaps the result of the comprehensive reorganisation process that the PLA has been undergoing for the past two and a half decades. Right from its birth, the PLA has been organised on regional lines. Inter-regional transfers of soldiers and junior ranks were not common, while the very senior commanders were rotated every two to three years (with a few exceptions) to prevent them from growing roots and building a 'following' among the ranks that might at some future date pose a threat to the regime. The PLA was organised into field armies and later group armies, the latter being more composite in nature. Today, however, the relatively new term "Military Area Command" (MAC) seems to be more popular. This implied that each of these regional armies was expected to operate in a largely self-contained fashion. As Kenneth Allen has noted, "the PLA did not go beyond developing a campaign strategy; a national military strategy was something the PLA was neither familiar nor comfortable". The 1991 Gulf War, however, plainly demonstrated the critical importance of employing all military resources at key points to ensure quick results if not outright victory. In such a scenario, even the relatively large PLA could not allow sizeable portions of its human and material resources remaining uncommitted. The 'War Zone' campaign, in short, indicates the belated realisation that to prosecute a modern war, the PLA would have to adopt 'Western' thinking and reorganise its logistics and other combat support elements so that its total offensive capability is brought to bear decisively, quickly and efficiently.

From the Principle of Mass to the Principle of Concentration of Firepower

This is a clear indication of the PLA's sincere attempts to ready itself for modern warfare. The term "mass" has always been synonymous with what is generally known as the principle of "concentration of force" at a time and place of own choosing. With rapid advances in firepower, especially that of air delivered precision weapons, it is no longer necessary to collect a huge force of men with equally large amounts of material before launching an attack. Today, it is possible to create decisive asymmetries with precision firepower either from ground or air. It is, thus, an efficient, quick and cost-effective way of winning a war without unnecessarily exposing troops to avoidable danger.

From the Four Single Service Campaign to Joint Campaigns

This is another indication of how far the PLA has come in its doctrinal and intellectual development. As noted earlier, throughout the Mao era, the PLA's focus had remained on the so-called superiority of men over weapons and, hence, the PLA played a dominant role in the country's defence along with the Second Artillery. The senior PRC leadership was also less familiar with the navy and the air force; the army, thus, also wielded considerable political influence. There were valid reasons for this state of affairs. First, the PLAAF did not possess suitable attack aircraft with adequate range and armament until the induction of the Su-27 fighters in the early 1990s. The huge and unwieldy combat air arm was largely primitive and its strategic strike bombers, the H-6 (Tu-16) was incapable of making any major difference except when used with nuclear weapons. The PLAN's role was also restricted to coastal defence and limited 'brown water' operations. This was for two reasons; first, a general absence of maritime tradition and second, its antiquated equipment. The concept of joint planning and operations was alien to the PLA top brass.

The other major reasons were inter-service rivalry and suspicions about the loyalty of the PLAAF commanders and men, who were reportedly close to Lin Biao, Mao's designated successor who, after a failed coup attempt, was in a mysterious air crash while fleeing the country. It was only after considerable personal efforts by Deng Xiaoping that the PLAAF slowly began to emerge from under this cloud of suspicion. Having learnt its lessons, the PLA is now slowly but definitely realising the need to integrate the other two services to fight a joint campaign. It was only in the latter half of the 1990s that the PLAAF commander was given the "responsibility for coordination" of major joint exercises; a belated realisation of the critical importance of the PLAN and PLAAF.

The Emergence of the RMA School

Before discussing the direction of the current doctrinal debate in China, it must be reiterated that the PRC has managed to compress two generations of military reforms in a mere fourteen years. At the end of the 1980s, the PLA was still largely equipped with the 1970s' technology weapon systems. Although the 'four modernisations' programme had begun in 1978, it was not until the beginning of the 1990s that the PRC gained the necessary economic strength and confidence to start translating the lessons of modern warfare into reality. "In early 1993, following a prolonged review of the Gulf War's 'lessons', the CMC called for two cardinal changes by the year 2000: change the military dependence on manpower and People's War to greater reliance on science and technology; switch plans for military preparedness from winning a conventional local war to winning a high-tech local war."[11]

The signing of the arms procurement protocols with Russia that underlined the long-term commitment to a defence relationship between the two countries must be seen as the true harbinger of change in the PLA's thinking and practice. Without Russian willingness to sell Su-27/30 advanced fighters, the Sovremenny class destroyers, Kilo class submarines, S10 and S 300 advance SAMs, RD-93 and Al-31F turbofan engines for the indigenous manufacture of the FC-1 and J-10 respectively, the PRC would have found it impossible to upgrade the PLA's weapon systems and the search for RMA technologies and the doctrine to go with them would have been merely wishful thinking. To be fair, it is not as if the PLA theorists did not appreciate the global trends in modern warfare in the 1980s and 1990s but China simply did not have the wherewithal to meet those needs nor was the PRC leadership very clear about what it exactly wanted its armed forces to do.

For the last decade or so, the adherents of the RMA school appear to be gaining some support, although according to Michael Pillsbury, the "people's war" and "local high-tech war" protagonists have not lost their predominance.[12] The RMA debate revolves around the fact that merely modernising the armed forces is not enough; the PLA must first develop a suitable doctrine to fit the fast developing technology and train its forces to wage what is commonly known as a "network centric war". There is also the talk a search for the magic weapon or the "system of systems" that would enable the PLA to vanquish the enemy in one fatal stroke. There is, thus, need for the PRC to first identify, acquire, operationalise new technologies and then train a sizeable portion of the PLA for effective war fighting in such an environment.

Having begun the process of modernisation in the three traditional branches of the PLA namely, the army, navy and the air force, the PRC theorists are now searching for ways to obtain capabilities in space and information warfare or what is generally referred to as the "fifth dimension", space being the fourth. The followers of this school are acutely aware of the role space-based weapon and combat support systems played in all the major war that the US fought in the last decade as also the critical need to develop a relatively invulnerable "command and control" apparatus to effectively employ these sophisticated systems. It is in these fields of high technology e.g. laser, directed energy, high-power microwave and anti-satellite weapons that can be launched from ground, air and space, that the search is apparently moving. China has been successful in obtaining the necessary assistance from many countries, including France, Italy, the UK, US, Israel and Russia. At the same time, China is also developing indigenous capabilities in all these fields as a hedge to possible arm-twisting or conditional sales of these systems in the future. By diversifying its search for dual-use technologies such as in telecommunications, commercial satellite launch and operations, industrial lasers, high-speed computing and data handling and building its industrial capacity to produce this hardware in the country, China has succeeded in large measure to avoid possible impediments or sanctions.

The other schools continue to recommend a slower and less dramatic change in the doctrine for two reasons. The "people's war" school largely populated by hardliners of the PLA, does not want to dilute the role of the army, while the "local high-tech war" school also composed of pragmatists do not want to expose the nation to a suddenly emerging threat in the Taiwan Strait that can only be faced with forces in being and not await the actual fruition of the RMA systems under development. The continued if somewhat less publicised build up of China's Short Range Ballistic Missiles (SRBM) capabilities is noteworthy in this context. The PRC and PLA leadership has demonstrated its willingness and determination to use these missiles to deter Taiwan from going the Unilateral Declaration of Independence (UDI) way. These missiles are an ideal weapon of deterrence and intimidation; can be manufactured at low cost without depending on foreign technological assistance; and finally, is indeed easy use from inside China. There are reports of China continuously improving the accuracy of these missiles by linking their terminal guidance tc an indigenously developed Global Positioning System (GPS) navigation suite. This, in short, is a hedge against an unforeseen contingency on China's periphery and India needs to take note of it.

Assessment

Having briefly seen the slow, almost grinding and even confusing evolution of the PLA doctrine over the last fifty-four years, it is evident that the hype about ideology was generally a camouflage for confusion and even inefficiency. Mao's word, no doubt, was gospel but even he was often indulging in double-speak simply because China neither had the money nor access to modern technology.[3] The frequent changes in the doctrinal debate also indicate some confusion in understanding the intricacies of modern warfare and, more importantly, the long suffering paranoia that the PRC leadership has suffered even when the external threat was minimal if not totally non-existent.

It also seems clear that there has been far too much interest in every turn and twist in the doctrinal debate when, in fact, the PRC was all along fighting its wars in the most pragmatic and flexible manner possible. While the most recent 'use of force', the missile firing exercises over the Taiwan Strait in 1995 and 1996, was entirely founded in Chinese concerns about the likely wayward behaviour of the Taiwanese government, its occupation of the Mischief Reef in 1995, the Nansha islands in 1988 or the Xisha in 1974 are all examples of pragmatism and expansion when the opportunity presented itself. The 1979 attack on Vietnam was born out of over-confidence in the PLA's fighting capabilities and an imagined slight to the PRC's political status in the eyes of the international community. Its disastrous performance in the war, however, did neither, but, in fact, showed China's weakness in bold relief. In the future too, China's military policy would most likely continue to be guided by the existing power balance vis-à-vis the adversary and its position in the global community with little reference to China's military doctrine. The Chinese have yet to master the art of modern high-tech warfare and merely obtaining sophisticated weapons may not be enough. What the PRC may, however, do is deftly employ its newly acquired weapon systems to intimidate, coerce or deter the potential foe without actually resorting to war. Should war become inevitable, the PRC would once again rely on its SRBMs and long reach firepower to intimidate the foe so that China can quickly extricate itself from the war without loss of face. It will be a long time before the PLA can truly absorb the much talked about RMA weapon systems and train its forces to gain the necessary confidence to operate them effectively. But even without perfecting or actually reaching a reasonable level of confidence in operating RMA technologies, the PRC can exert considerable military influence in the Asia pacific region. Last but not least, the history of the PLA shows that the political leadership has often interfered in the decision making at the field level. The PRC leadership may not be able to do that in the future because the essence of RMA is greater decentralisation and autonomy to field commanders. Indeed, if the PRC leadership decides to interfere in military decision-making it may well end up eroding the effectiveness of these weapon systems. Finally, as Lewis succinctly puts it, "Mao's revolutionary doctrine, if not ancient strategies, long ago dictated the *threatened use of force in manipulating an adversary's responses,* and China's leaders have consistently demanded the military's acquiescence to political authority when calibrating the magnitude and timing of the pain, if any, to be inflicted. In these circumstances, recourse to force always remains subordinate to political stipulations that can *violate standard military principles*"[13] (emphasis added).

Notes

1 John Wilson Lewis & Xue Litai, *China's Strategic Seapower* (Stanford, CA, Stanford Press, 1993) p.

2 I am indebted to Air Cmde Jasjit Singh for bringing the critical importance of this aspect in understanding China's military psyche and political mindset.

3 Nan Li, "The PLA's Evolving War-fighting Doctrine, Strategy and Tactics, 1985-95: Chinese Perspective", in David Shambaugh and Richard Yang (ed)., *China's Military in Transition* (Oxford: Clarendon Press, 1997), p. 179.

4 John Wilson Lewis, "China's Search for a Modern Air Force", *International Security,* Vol. 24, No. 1(Summer 1999), p. 73.

5 Ibid. p. 70.

6 Nan Li, "The PLA's Evolving Campaign Doctrine and Strategies", in Mulvenon & Richard H Yang, *The People's Liberation Army in the Information Age* (Santa Monica: RAND, Conference Proceedings, 1999), p. 129.

7 Lewis, n. 1, p. 216. All quotes in this paragraph are from Lewis.

8 Ibid, in the reference note on page 318.

9 David M Finkelstein, "China's National Military Strategy", in James C. Mulvenon & Richard H. Yang, n. 6.

10 Lewis, n. 4, p. 89.

11 Ibid., p. 84.

12 *Michael Pillsbury,* ed., *Chinese Views of Future Warfare* (Washington, DC: National Defence University Press, 1997).

13 Lewis, n. 4, p. 86.

4

Preparing the Army for the Future

Introduction

In general usage, the term PLA denotes the PRC's armed forces and includes the navy, air force and second artillery or its strategic nuclear and missile forces. "The PLA was established on August 1, 1927, and consisted of land forces only in its early days."[1] The PLA actually traces its origins to 01 August, 1927 because it was on this day that the Northern Expeditionary Army was formed. With the formation of the First Workers and Peasants Revolutionary Army on 11 September, 1927, the PLA was fast gathering more adherents. Zhou Enlai and Mao Zedong were the leaders of these two armies respectively.[2] As noted earlier, this rag-tag force slowly acquired momentum till in 1949, it emerged victorious in the long civil war. The PLA had played a significant role in fighting the Japanese invasion and in the liberation of the country from imperialist and feudal elements and hence its importance in political and strategic decision-making was axiomatic.

Given the fact that the revolutionary leadership held the true levers of power by occupying the highest positions in the party and the army, a 'double-hatting' approach which continues to this day, was adopted. In this scheme of things, the senior CCP functionaries also held senior positions in the

PLA. This often made it extremely difficult to distinguish between the areas of responsibility of the party and the army. To be sure, Mao and other senior leaders repeatedly highlighted the fact that the 'Party' always controlled the gun. Civilian control of the PLA is, thus, a long and enduring tradition. Even today, the PLA owes its allegiance to the CPC and through it to the state. The fact that the CMC always has at least two or three senior PLA generals as its members means that the PLA also always has an influencial voice in the national decision making. In the recent past, however, its presence in the highest political and party bodies has been dwindling somewhat.

We have already seen the stellar role that the PLA played during the Korean War, China's first confrontation with a super power. Its performance during the war enhanced its importance although General Teng Dehuai was replaced by Lin Biao after the first successful counter-attack that pushed the American forces back to the 38th Parallel. As noted earlier, throughout the early 1950s, China continued to entertain fears of an imminent attack by the US backed nationalist forces on the island of Formosa now called Taiwan, in collusion with sympathisers, separatists and other anti-Communist elements on the mainland. This fear further reinforced the need to maintain a strong defence, which naturally and logically only enhanced PLA's prominence in PRC's national affairs.

Strength

At the time of the founding of the PRC in 1949, the strength of the PLA which mostly comprised only the ground forces, was indeed very large since apparently all those Party cadres who had participated in the forty-year long civil war had to be accommodated and given some employment. During the process of state and government formation, many of the uniformed members were transferred to other civilian duties and, hence, the strength of the PLA kept reducing.

There were attempts to downsize the PLA, especially its ground elements even during the Mao era.

"At a landmark meeting in May-June 1985, the Central Military Commission (CMC) approved Deng's call for a phased reduction of one million troops and use of the resulting savings for economic development."[3] This was in the face of domestic resistance which Deng had expected but he gradually overcame it. Today its strength stands at 2,270,000 including perhaps 1000,000 conscripts and 136,000 women. The term of service for selective conscription in all services is two years.[4] Of these, the army component stands at an estimated 1,600,000, including some 800,000 conscripts. In the recent past, the PRC leadership has reduced the strength of the PLA by over a million and half soldiers. The modernisation programme cannot obviously equip such a large force with modern weapons and equipment and, therefore, the PLA is developing what is generally called "islands of excellence" by selectively upgrading the training and equipment of its ground forces.

Organisation

Given China's huge continental geographic extent, long land borders, difficult mountainous terrain and the consequent logistical problems, right from the beginning, the PLA has been organised on a regional basis. At first, the country was divided into 13 military regions (MR), which were reduced to 11 in the 1970s and later to seven. This reduction indicates a gradual improvement in surface communications and logistic support. Recruitment and postings excepting those of very senior ranks were thus restricted to the region and readily provided a sizeable number of trained and acclimatised troops to meet any challenge to the country's security. Regional organisation also ensured that the soldiers were fighting to defend their homes and since the national level party organisation was also mirrored down to the provincial and district levels, it made for better and closer political control of troops. Today, the seven MRs have many military districts under them and the term MR is gradually giving way to military area commands (MAC), once again denoting the PLA thinking on flexibility in joint operations. As stated earlier, selective modernisation would naturally call for some if not sizeable movement of forces in an external or internal threat scenario.

The PLA, in its formative years comprised 'field armies' which have in the recent past been downsized and reorganised into combined or integrated group armies (GA) and currently number 21. The GA comprises elements of all arms but the exact composition depends on the specific needs of the MAC. There are 44 infantry divisions, including five mechanical infantry and 24 motor infantry. Three of these divisions have a national level rapid reaction role and at least nine others are with a regional level rapid reaction role and are ready to mobilise in 24-48 hours. In addition, there are nine armoured divisions, 12 armoured brigades, one mechanical infantry, 22 motor infantry, six artillery divisions, 15 artillery brigades, one Anti-tank Brigade, eight 'Surface to Air Missiles'/ Anti Aircraft-Artillery (SAM/AAA) brigades, 13 AAA brigades and three army aviation regiments.

Further, independent combat units include five infantry divisions, and one infantry, two armoured, one motor infantry, one artillery, three anti-tank, and one SSM brigades and ten army aviation regiments. In addition, there re some 12 infantry divisions, four armoured and one mountain brigades and 87 infantry regiments as part of local forces that include garrison, border and coastal duties units.

GA strength varies from 30-60,000 troops and a GA usually comprises two to three mechanised/motorised infantry divisions/brigade, one armoured division/brigade, one artillery division/brigade, one SAM /AAA or AAA brigade, with their combat readiness category varying, with 10 GA at Category A and 11 at Category B. Reorganisation from the divisional to the brigade structure has been in progress for the last decade or so.[5] This reorganisation is once again aimed at improving mobility, command and control, joint operations and rapid reaction.

Deployment

Presently, the deployment of the 21 GA units is as follows:

- **Northeast** Shenyang MR (Heilongjiang, Jilin, Liaoning MD): Estimated strength of 250,000 with 4 GA; 2 armd, 2 mech, 3 mot, 1 arty div; 5 mot, 3 art div; 2 armd, 5 mot, 3 art 1 SAM/AAA, 3AAA and 1 ATK bde.
- **North** Beijing MR (Beijing, Tianjin Garrison, Nei Mongol, Hebei, Shanxi MD): Estimated strength of 300,000 troops with 5GA; 2 armd, 1mech, 5motor, 1 arty, div; 3 armd, 7motor inf, 4 arty, 2 SAM/AAA, 3 AAA bde; 1 ATK regt.

Modernisation

- **West** Lanzhou MR (Ningxia, Shaanxi, Gansu, Qing-hai, Xinjiang, South Xinjiang MD): Estimated strength of 220,000 with 2GA; 1 armd, 2 motor inf div; 1 armd, 2motor inf, 2 arty, 2AAA bde; 1 ATK regt.
- **Southwest** Chengdu MR (inclding Chongqing Garrison, Sichuan, Yunnan, Xizang MD): Estimated strength of 180,000 troops with 2 GA; 4 motor inf, 1 arty div; 2 armd, 1 arty, 2 AAA bde.
- **South** Guangzhou MR (Hubei, Hunan, Guangdong, Guangxi, Hainan, MD): Estimated strength of 180,000 troops with 2 GA; 4 motor inf, 1 arty div; 2 armd, 1 arty, 1 SAM/AAA, 1 AAA bde. In addition there are an estimated 7000 troops in Hong Kong with 1inf bde including 3 inf, 1 mech inf, 1 arty, regt and 1 engr bn.

- **Centre** Jian MR (Shandong, Henan MD): Estimated strength of 190,000 troops with 3 GA; 2 armd, 1 mech inf, 3 motor inf, 1 arty div; 1 armd, 1 mech inf, 4 motor inf, 2 arty, 2SAM/AAA1 AAA de.
- **East** Nanjing MR (Shanghai Garrison, Jiangu, Zhejiang, Fujian, Jiangxi, Anhui MD): Estimated strength of 250,000 troops with 3 GA; 2 armd, 1 mech inf, 3 motor inf 1 arty div; 1 armd, 4 motor inf, 2 arty, 2 SAM/AAA, 1 AAA bde and 1 ATK regt.[6]

Equipment and Hardware

The PLA Army has some 7,000 main battle tanks and 1,200 light tanks, over 5000 APC/AFV, 14,000 towed artillary pieces, 1,200 SP arty 2,500 MRL and numerous mortars and other standard equipment. It also has a number of SSM of HY-2/C-201/CSS-3, HY-4/C-401/CSS-7/M-11 missiles. The PLA's armoured force comprises largely old and antiquated tanks such as the Type59-I/II while the 500 Type-79, 900 Type-88A/B, 550 Type 88-C, 60 Type-98 and some Type-96 make up the relatively modern equipment. It would be evident that the PRC leadership is only slowly modernising its ground forces since there is literally no ground threat. The only real reasons for maintaining such huge forces are deterrence and for possible internal contingencies. The corps-strong airborne force is also given added mobility for internal duties and was used during the 1989 Tiananmen Square disturbances. The PLA knows full well that it needs to quickly improve its logistics and combined-arms capabilities. So far the army has always enjoyed a position of preeminence. Since the 1991 Gulf War and the 1999 Kosovo conflict, however, the army has realised that joint warfare capabilities deserve greater and more urgent attention. It is currently trying to improve its combat-worthiness as well as give the necessary importance to specialist services such as the signals and engineers, command and control and training to fight alongside the navy and air force.

Modernisation and Restructuring

With the signing of the 14 February, 1950 Friendship Treaty with the Soviet Union, China had begun to receive Soviet arms, equipment, ships, aircraft as well as technological assistance to build its own defence industry. The PLA, thus, developed on the lines of Soviet armed forces and by the end of the first decade, had grown into a formidable armed force. Sino-Soviet tensions of the early 1960s, however, adversely affected the development and modernisation of the PLA which continued to operate obsolete equipment well into the 1980s. It was only after Deng Xiaoping's economic reforms started to make significant progress in the mid 1980s that the PLA could think of demanding funds for its modernisation.

Due perhaps to its origins and its close links with the CPC, the PLA of the 1950s grew without any strict rules and regulations concerning promotions, transfers, training and professional competence and as a result, there were many cases of subordinates outranking their commanding officers. Frequent purges and transfers on account of domestic political upheavals also affected the PLA. Interestingly, during the nearly ten year long disastrous Cultural Revolution, the PLA leadership remained loyal to the party and yet was used to suppress radical elements and Red Guards.

The PLA has been subjected to restructuring and reforms almost throughout its post-1949 history. "Since the early 1990s, the PLA has accelerated the pace of its modernisation." The significant feature of this process is that along with its arms and equipment, the PLA is also addressing its doctrinal deficiencies." PLA restructuring, as noted earlier was obstructed by lack of funds due to an underdeveloped national economy and an "obsolete notion that the country's national defence was fundamentally continental."[7] More importantly, the PLA was created as a ground force. Specialised services appeared in its force structure only after the army had achieved absolute control over all of the armed forces. Both air force and navy headquarters were not established until after the PLA had won the national civil war. As a result, infantry men have long dominated the top brass and they have the best chance of being promoted to the CMC. Domination by infantry officers has inhibited the restructuring of the PLA to suit modern warfare, which inevitably calls for an enhanced role for specialised services. "Since the late 1980s rapid economic growth has led to unprecedented financial allocations for the modernisation of the military. With the military budget doubling every four to five years, the PLA is now able to undertake long-term development projects that are costly and technologically demanding."[8]

"Two elements have been crucial in guiding the PLA's restructuring efforts: new military thinking and the gradual introduction of advanced hardware". As mentioned earlier, "the PLA has increasingly embraced the idea that a high-tech war is by nature an offensive war." Since the specialised services are regarded as the main vehicles for offensive operations, they have gained a special status. Moreover, the PLA's projected defence in depth will eventually give the specialised services a leading role, especially to those that can operate into international air and maritime space. According to Lieutenant General Xu Fangting, "the PLA's future war actions will take the form of united campaigns in which the Air Force, Navy, and Strategic Missile Force play a decisive role."[9]

Different defence strategies catering to different threat perceptions have an important bearing on projected force levels. After 1974, when Mao Zedong entrusted Deng Xiaoping with the everyday affairs of the CMC, Deng initiated two major rounds of force reduction, each time as the result of a changed focus in the national defence strategy. In 1975, the CMC concluded that a massive Soviet invasion was more of a myth than reality. With Mao's support and Deng's guidance, the CMC implemented a three-year plan of force reduction, cutting the overall force level of the PLA from 6.1 million to 4.5 million, making it the largest reduction in the history of the PLA since the PRC was established in 1949. The army was reduced by 32 per cent, the strategic missile force by 27.2 per cent, and the air force by 13.4 per cent. The staffs of the headquarters from the CMC to the eleven MACs were trimmed by 32.2 per cent. Only the navy increased it number of sailors. Within the ground force, field army infantrymen were cut back by 28.6 per cent, border troops by 18.3 per cent; local garrisons by 34.5 per cent and artillery units by 10 per cent. The deepest reduction took place in the engineering corps and the railway corps, the former being cut by 62 per cent and the latter by 65 per cent. This line of thinking also informed the PLA's reduction a decade later when the emphasis shifted to fighting a flexible limited regional war, which was thought to involve no large-scale military campaigns, but instead required a quick and effective response to actions along China's land boarders. As a result one million soldiers were demobilised in the mid-1980s.[10]

Restructuring plans were put on hold following the June 1989 Tiananmen Square incidents during which the army had to be called out to quell student unrest. The PRC leadership took time to ensure that the PLA continued to support the CPC's decisions and that there was no major chance of the trouble recurring in the near future. It would have been foolish to effect major cuts in the PLA when the Party needed its support to maintain internal stability. Many top departments were merged and their functions streamlined. As quoted in You Ji, a senior PLA researcher believed that, "the rationale for the merger was that the PLA central command had over 300 specialised agencies at the divisional level or above. If each of them issued only one instruction a year to the combat units at lower levels, it would still make the running of these units difficult because it meant that these units received about one instruction a day, instructions they have to implement and be prepared for inspection. This is not to mention the fact that many of the instructions are overlapping and confusing." It is easy to imagine the sheer magnitude and complexity of modernising an army that had been run on party lines for over five decades.

The Military Rank System

The history of the military rank system in the PLA makes interesting reading. Writing in 1996 Srikant Kondapally, an Indian researcher with the IDSA says, "The recent emphasis in the PRC on professionalism of the armed forces once again brings to the fore the issue of reintroducing ranks in the PLA". On the system of instituting ranks in the PLA and its predecessor the Chinese Workers and Peasants Red Army, the CCP leaders differed and pursued their respective policies. The attempts at creating a rank system in the armed forces during the pre-Liberation period proved to be a failure. In 1951, a system of 23 grades was instituted. However, by 1955, difficulties in implementing this grade system were detected. In the 1951 grade system, there was a wide gap of two grades between the ranks and the authority attached to these posts. As this could result in an improper exercise of authority, changes were sought to be made in 1955. On February 8, 1955, the Standing Committee of the National People's Congress (NPC) passed a resolution for establishing the rank system and even implemented this project for over a decade. By this policy Zhu De, Chen Yi, He Long, Xu Xiangqian, Lin Biao, Liu Bocheng, Luo Ronghuan, Nie Rongzhen, Peng Dehuai, and Ye Jianying became the Ten Marshals of the PLA, along with two Generalissimos, 57 Generals and 1,359 Major Generals. This system was abolished on 22 May, 1965, with the onset of the Cultural Revolution. Recently, this project was taken up again in May 1984, but become effective from July 1, 1988".[11]

Command & Control

As noted before, the PLA reports to the CCP and through it to the central government. It is, therefore, only natural that high ranking PLA officials work at the pleasure of the Party general secretary and assist him in high level decision making as members of the CMC. On the question of PLA links to the CPC and its influence in politics it is safe to assume that neither will the PLA jeopardise its symbiotic relationship with the Party nor would the Party ask the PLA to act in a manner that would appear against the interests of the regime in particular and that of Chinese society in general. In overall terms, it is often surmised that the PLA will continue to wield considerable influence in the foreign policy arena but with the winds of economic reforms blowing stronger every day, it is also possible that the PLA would gradually lose some of its influence in high policy. The number of PLA generals on the CMC has been coming down in recent years. With the installation of young technocrats in the recent transfer of power, it is possible that the PLA leadership would also encourage more and more professionals to rise to higher ranks. In any case, with the gradual but definite move towards greater professionalism in the PLA, the chances of major disagreements between the civilian and military leadership would reduce simply because the PLA officers would not have the time and opportunity to play politics. Each side would most likely devise a workable equation except when the PLA perceives that the CPC leadership is taking anti-people measures that might give rise to Tiananmen type protests and force the PLA to take sides or to crack down on innocents. In all probability, the CCP and the PLA, through mutual consultation, try their utmost to maintain reasonable internal stability in China. The role and the position of the political commissar, a Party official paralleling every PLA post is thus not antithetical to the system. As Jiang Zemin has said, the PLA must at all times remain completely loyal to the CCP and, to that extent, the salience of political education would also remain paramount.

Such relationships may however, make it more difficult for the PLA to transform itself into a modern professional army. This is so because political interference is not compatible with professional thinking and discipline. For example, undue interference by Party officials may adversely affect military decision making. "Ideological and political purity, therefore, takes precedence over military training and readiness. The senior Chinese leadership perceives this as a prerequisite for stability."

Additionally, all military developments must be framed within the ideological bounds of the military thinking of Mao Zedong as further refined by Deng Xiaoping's peace time army building ideology and Jiang Zemin's "five general requirements". In effect, this means that debate and the development of doctrine and training must be justified in Mao's overall rubric of "people's war" and the military strategy of "active defence. This is precisely why the PLA may take a long time to become a truly professional army.

Assessment

From all accounts, it seems inevitable that the PLA is going through a fundamental systemic change towards greater professionalism. In the last two decades it has been downsized considerably even if the majority of the retrenched soldiers were either absorbed in the newly formed People's Armed Police (PAP) and/or transferred to the railways and other civilian departments where they rightly belonged in the first place. The PLA has thus shed a large portion of its purely civilian responsibilities. The process of regularising includes making the necessary laws, rules and regulations that would ensure transparent and just terms and conditions of service, fair promotions, opportunities for career progression and a relatively trouble-free administration of this huge force. Restructuring and reorganisation have together attempted to ease the difficulties of joint operations and given the specialists within the army such as signals, engineers, armour, and artillery, and air defence elements as also the navy and air force considerable voice in central decision making. It is, however, too early to predict if the other two services have indeed broken out of the army/infantry straitjacket.

The PLA officers are generally better educated and are becoming increasingly technology savvy. The insistence on compulsory retirement at fixed ages and promotions only to those who have successfully completed higher staff training at the prestigious National Defence University (NDU) are indicators of rapid professionalism in the army. The Defence White Paper stresses that the small increases in the country's defence budget are essentially to improve the standard of living of the PLA. According to an earlier edict, the PLA officer was required to be paid a salary that was 20 per cent higher than his civilian counterpart. This differential has been eroding in recent times due to, firstly, the ban on the PLA to run enterprises; and, secondly, because of the prevalence of corruption in civilian government sectors. As a result, there are not too many young and able students ready and eager to join the PLA. "Young and bright students would rather opt for higher education and proceed abroad for higher studies", a PLA officer told this author. It seems that, like in many other countries, China too faces many problems of recruitment and retention made the worse by selective conscription in which the wards of senior and influential Party officials usually succeed in dodging the draft even when physically fit and suitable for recruitment.

As we have seen, the emphasis is on selective modernisation of the PLA with only the rapid reaction or 'fist' units getting a lion's share of additional funds for modern equipment and hardware. The objective according to the latest issue of IISS Military Balance appears to be to reduce the number of divisions from 100 to 50 over the next few years.

Even so, the PLA's contribution to the PRC's power projection capabilities would remain minimal as China is unlikely to either wage a major continental war or station its forces too far away from its shores. The current emphasis on joint planning and joint operations is, in fact, reducing the usually high importance given to the PLA. The 1998 ban on the PLA running enterprises is also a right step in making the PLA a truly professional army.

Notes

1 China's Defence White Paper-December 2002; available on the internet at <fbis.dod.gov>

2 Bajwa, *Modernisation of the PLA* (New Delhi: Lancers, 2002), Appendix A, p. 419.

3 John Wilson Lewis and Xue Litai, *China's Strategic Sea Power* (Stanford: Stanford University Press, 1994), p. 214.

4 *IISS Military Balance*, 2002-03, pp. 145-148.

5 Ibid, p. 145.

6 Ibid. p. 146

7 You-Ji, *The Armed Forces of China* (London: IB Tauris Publishers, 1999), p.29

8 Ibid., p. 29

9 Ibid., p.31

10 Ibid., p. 33.

11 Srikant Kondapalli, *The People's Liberation Army: Evolving Dynamics* (New Delhi: IDSA, 1996), pp.40-41.

5

From Coastal Defence to Blue Waters

"The present trend of development is towards guided missiles, electronics, nuclear, capability and automation, to be gradually achieved through repairing and refitting old vessels, purchase from abroad, transferred technologies and improving on existing models."

—Adm Zhang Lianzhong
PLAN commander, 1991[1]

Introduction

The 2002 PRC Defence White Paper describes the role and the functions of the PLA Navy as follows:

The Navy of the PLA was established on April 23, 1949. Its primary missions are, independently or jointly with the Army and Air Force, to guard against enemy invasion from the sea, defend the state's sovereignty over its territorial waters, and safeguard the state's maritime rights and interests. The Navy has such arms as the submarine, surface, naval aviation, coastal defence and Marine Corps, as well as other specialized units.

Under the Navy, there are three fleets, namely, the Beihai, Donghai and Nanhai fleets, as well as the Naval Aviation Department. Each fleet has bases, maritime garrison commands, flotillas and squadrons under its command. The submarine force is composed both of conventional and nuclear-powered units, with underwater attack and some nuclear counterattack capabilities respectively. The nuclear-powered submarine force, which assumes the strategic nuclear counterattack mission, is under the direct command of the CMC. The surface force has combat and support units, which have anti-ship, anti-submarine, air defence, mine warfare and shore attack capabilities.

The naval aviation is composed of bomber, fighter-bomber, attacker, fighter, anti-submarine and reconnaissance units, and security, ECM, transport, rescue and air refuelling units, which have reconnaissance, security, anti-ship, anti-submarine and air defence capabilities. The organisational order is: naval aviation department, fleet aviation, and aviation division and regiment.

The naval coastal defence force is composed of shore-to-ship missile and coastal artillery units, which have capabilities to defend China's coasts. The marine corps has infantry, artillery, armour and engineering units, as well as reconnaissance, chemical defence and communications units. It is a rapid assault force for amphibious operations.[2]

Writing in 1997, David Shambaugh agrees, "The key to power projection is, of course, the PLA Navy (PLAN). Without a blue water capability, forces cannot be despatched out of area, much less re-supplied, and sustained, in any significant numbers. Over the past decade, the PLAN has moved from a brown-water coastal to green-water capability, with some vessels being blue-water capable. The PLAN has sent flotillas to and through the Indian Ocean, Western Pacific and Arctic seas in recent years. But basically only one 18-ship task force can remain at sea for any sustained period without forward operating bases"[3]

"In aggregate terms, the PLAN is the third largest in the world. It is roughly equal in size (but not in armament) to the Russian Pacific Fleet. However in qualitative terms, when measured against other navies in Asia, one finds that the Japanese Self-Defence Force, the Indian Navy, the combined naval forces of the ASEAN nations and the elements of Taiwan's Navy are all superior to the PLAN."[4] This was the assessment in 1995. The PLAN has since then inducted many new ships and more importantly, proposes to upgrade its overall offensive capabilities and reach in a significant manner in the near future.

Strength

According to the latest issue of the *IISS Military Balance*, an authoritative source for military information, the PLAN has on its strength the following ships, submarines and aircraft. With a strength of some 250,000 personnel including coastal regional defence forces and 26,000 naval aviation, some 10,000 marines and 40,000 conscripts the PLAN has already shed some 10,000 personnel in the last five years or so. Its present order of battle comprises 1 strategic nuclear submarine, 67 tactical submarines, 21 destroyers, some 42 frigates, about 368 patrol and coastal combatants and 39 marine warfare craft, 56 amphibious landing craft and about 163 support and miscellaneous crafts and some 472 shore-based combat aircraft and 45 armed helicopters of its naval aviation.

Deployment

North Sea Fleet. This fleet is responsible for the coastal defence from the border of the Democaratic People's Republic of Korea (DPRK) (Yalu River) to south of Linyungang and equates with Shenyang, Beijing and Jinan Military Region (MR), and to seaword. Its bases include the Fleet HQ at Qingdao, Dalian, Huludao, Weihai, Chengshan,Yuchi and nine coastal districts.

East Sea Fleet. This fleet is responsible for the coastal defence from south of Lianyungang to Dongshan and equates to Nanjing MR and to seaward. Its bases include the Fleet

HQ at Dongqian Lake (Ninbo), Shanghai Naval Base, Dinghai, Hangzhou, Xiangshan, and seven coastal districts.

South Sea Fleet. This fleet is responsible for the coastal defence from Dongshan to the Vietnam border and equates to Guangzhou MR and to seaward including Paracel and Spratly Islands in the South China Sea. Its bases include Hong Kong, Yulin and Guangzhou.

Evolving Role and Strategy

Although the PLAN came into being before the birth of the PRC, due to a very strong bias in favour of a continental strategy the Chinese leadership did not really nurture the maritime tradition of the Ming era. It was during the latter part of the Ming dynasty (1368-1644) that Chinese mariners had sailed as far as the coast of Africa and had also touched India's Malabar coast where even today Chinese fishing nets are in daily use. It is somewhat surprising that even though most if not all of exploitation of Chinese weaknesses by foreign powers had been through the sea the Chinese continued to place excessive concern in the continental threat from the north and the west rather than from the seaward. As a result, the PLAN too suffered a general neglect that was remedied only in the post 1978 military modernisation programme.

The PRC had, in fact, employed its rather limited naval power in many conflicts including Korea, Taiwan Strait crisis of the 1950s, the occupation of Xisha Islands in 1974, and Nansha in 1988 and most recently of Mischief Reef in 1995. The PLAN had performed reasonably well but it was perhaps due the influence of the infantry generals at the helm of affairs that the PLAN was seen only as a coastal defence force adequate to meet the seaward threat to the PRC. Even so, the PLAN had over the years built up a formidable force of some 50 submarines and an equal number of surface combatants. Most if not all of these ships and submarines were of Soviet origin and were not exactly modern by current global standards but were nevertheless more than adequate to deter a major invasion of the Chinese mainland. As we shall see in the next section the PRC's air arm also remained antiquated during most of the same period but was nevertheless maintained as an instrument of deterrence. Whether or not the PRC in fact deterred a major threat to its territorial integrity because of its possession of these obsolescent weapons systems is a moot point. But with Soviet help, the PRC had built a formidable capacity for ship building over time. "China State Shipbuilding Corporation operates 27 shipyards, 66 manufacturing plants and 37 research and development centres for both commercial and naval vessels."[5] Another reason for the continued neglect of the navy was a serious paucity of funds until after the success of the economic reforms programme in the 1980s.

The PRC leadership, however, seems to have been quite conscious of the role of sea based nuclear capability in China's overall nuclear posture of deterrence. This is amply proved by the fact that even when the PRC's economy was not doing well funding for the indigenous nuclear submarine programme was more or less sustained till the Chinese scientists finally built a nuclear submarine and a Submarine Launched Ballistic Missile (SLBM). It is noteworthy that Mao who was normally critical of the efficacy of nuclear weapons gave his approval for the launch of the nuclear submarine programme as early as in 1958 and chose Gen. Nei Rongzhen, an able leader as its head.[6] That it took a long and arduous 23 years for the programme to fructify is proof of Chinese determination and perseverance considering the numerous hurdles and difficulties the programme faced. Even today, the only strategic submarine built under this programme is reportedly plagued with many serious problems and is perhaps not easy to maintain, but given Chinese ambiguity and secrecy, no country can afford to dismiss the PRC's sea-based nuclear capability of the triad.

We should now turn our attention to China's renewed interest in the peace time building of its naval assets when it faced no objective external threat. Was it simply part of the overall military modernisation drive or was it specifically aimed at building a naval capability that was not restricted to defensive needs alone? Since with active Soviet assistance the PLAN had already built up a formidable arsenal of ships, submarines and weapons, the ongoing modernisation, it is evident, is aimed at replacing old equipment with state-of-the-art systems that would also give the PRC the capability to project power abroad. As we have seen before, the PRC strategists were concerned with the possibility of an all-out attack by a superpower. In the north, the Soviet threat loomed large, especially after the Soviet invasion of Czechoslovakia in 1968 and the fear of the Brezhnev Doctrine of "limited sovereignty' being applied to China. These fears were further magnified with border tensions escalating in early 1969. As a result, the PRC leadership had to prepare itself to face a threat of a major war which would be fought under the overhang of nuclear deterrence, such as it existed at that time, using the hallowed 'people's war' doctrine simply because the PLA was not technologically ready to fight a modern war. As noted before, Mao recommended the 'people's war' doctrine in his public utterances to ensure popular support for any future war but was secretly demanding, "What they have, we must also have". The role of the navy was thus, limited.

The other threat that China faced was that of US intervention in the event of a war over Taiwan. This threat would have materialised from the seaward but even here the PLA generals saw only a limited supporting role for the PLAN: one of obstructing the invaders and defending the coastline until the PLA's ground forces were ready to deal with the invasion proper. Once again, the stress would be on luring the enemy deep, yielding ground if needed, and then aiming for ultimate victory in a protracted war. "Stretched over time the doctrine of active defence in fact meant protracted war."[7] Curiously, the PLA strategists remained worried about the threat of a sharp border war with China's neighbours. "They deemed Vietnam and India to be the most probable future adversaries"[8] PLA naval strategy, thus, developed between two extremes; first, strategic nuclear capability conferred by the submarine based 'survivable deterrent' and second, coastal defence which extended to safeguarding China's coastal targets against the invasion of the mainland and a limited capability to blockade Taiwan.

The PLAN's current thinking, however, is fast following what one may call 'standard western naval strategy' of gradually building its capabilities so as to extend its reach from immediate coastal and off-shore defence to at least the first chain of islands, that would cover the Paracel and Spratley islands, claimed by China. This amounts to the beginning of its power projection capabilities in the South China Sea and the approaches to the Pacific in the East and the Indian Ocean in the West. Terms such as 'sea denial' or 'sea control' are, however, not used in Chinese writings but when Deng Xiaoping said that the PLAN was capable of blockading Taiwan he perhaps meant that the Chinese sea power was capable of denying the free use of the sea to any forces coming to Taiwan's help.

Modernisation

Since the 1980s, and particularly after the 1979 disastrous border war with Vietnam, the PLA's weapons procurements have been driven by the doctrinal shift from fighting a 'people's war under modern conditions' to a local/border war under high-tech conditions. The lessons of the first Gulf War of 1991 also seem to have deeply influenced Chinese thinking on modern warfare. Once the threat of an all-out attack by the superpowers disappeared what remained was the threat of a limited war with neighbours and to win this short-duration, local war China had no option but to procure modern weapons that would give it a fair chance of success. As noted above, the real aim of the PLAN's modernisation is to extend its reach.

From 1985 to 1995 the pace of acquisitions and modernisation was slow, due mainly to a general paucity of funds. Even so, at least two destroyers, nine frigates and three submarines and a host of other support and supply ships had been added.

Submarines

Other than the single Xia class SSBN with 12 JL-2 SLBMs, a sea launched version of the DF-21 IRBMs, the PLAN has five Han class nuclear powered submarines. It was reported that the Chinese faced some problems with these submarines when the first of these five was launched in the mid 1970s but the "later models have been improved with the most recent capable of launching C-801or YJ-6 SSM, a derivative of the French Exocet, from the surface."[9] Interestingly, the Chinese have developed a diesel-electric submarine, the Song (Type 039 similar to the French Agosta 90B in external shape), rather than continue the development of the Han, perhaps confirming the reliability and ease of operations of the diesel powered submarines as against the Han which is nuclear powered. In addition, in 1995, the PRC ordered four Kilo class (2 Type-EKM 877 and 2 Type- EKM 636) submarines from Russia. According to one report, "In May this year (2002), China ordered eight Kilo 636 submarines armed with Club-S surface-to-surface missiles (SSMs) from Russia. It is planned that the submarines will be delivered to China within five years. Because of this, Russia is planning to produce these submarines simultaneously in different shipyards. As in the case in the aerospace industry, China will not be giving up the Song-class conventional submarine-upgrading project."[10]

These are the latest versions of the submarines operated by the Russian Navy and all four have been inducted in the service. The indigenously designed, "Ming (Type-035) had also met with a number of difficulties and its production was suspended for a while in the mid 1980s." If China undertakes licensed production of the latest Kilo, the Ming may also be upgraded with new Russian technology; 16 of which have already been improved. Of the 84 Romeos, only some 35 remain but these are being slowly phased out.[11]

Destroyers

The latest in the acquisition programme are the Sovremenny class destroyers that China has received from Russia. Named the Hangzhou, two are already with the PLAN while two more are yet to join the service. The development of the Luhai class indigenously built destroyers has reportedly been plagued with "propulsion system problems". The Chinese have acknowledged boiler leak accidents on long distance voyages. Since the General Electric LM-2500 turbine engines are no longer available, the PLA Navy is using turbine engines GTE80 (DN80) from Ukraine."[12] Two other recent additions are the Luhu which began its sea trials in 1992 and the improved Luda. In 1997, Shambaugh said, "The PLAN's destroyers are undergoing a generational change. The 15 Luda class destroyers (Type 051) continue to be the principal surface combatants, but the Dalian and Jiangnan shipyards have begun to turn out an upgraded Luda *III* and the first Luhu class (Type 052) destroyers, respectively." Today, however, the PLAN already has 1 Luhai, 2 Luhu, 1 Luda *III*, 2 modified Luda and 13 other Luda destroyers in addition to the Sovremennys, giving it a considerable reach for force if not projection. The Luhu is powered with two of the five General Electric LM-2500 turbine engines acquired from the United States before 1989 and has the French Crotale SAM, the French Exocet-based C-801 ship-to-ship missile (SSM) Italian torpedo launchers and the two French Dauphin-2 based Z-9A helicopters that it carries on board. It is thus clearly seen that the PRC is following its oft declared policy of selective foreign purchases supplemented with robust indigenous development with foreign technology.

Frigates

In this department, the PLAN possesses some 30 Jianghu in three variants, 8 Jiangwei *II* and 4 Jiangwei *I*. "The Jianghu types have been undergoing modernisation programmes utilising imported French systems or French/European systems built under licence in China. Major improvements include C-801 launchers, improved missile and gun fire control electronic suites, a towed variable-depth sonar system and improved torpedo capabilities."[13] It was thought that the PLAN would by the year 2000 have some six Jiangwei frigates in service, but the actual number is exactly double,12, an indication of China's accelerated naval building.

Amphibious and Supply Capability

In addition to increasing the strength and the quality of the major combatants, the PLAN is also receiving attention in the area of combat support and amphibious ships. The PLAN has many more oilers, store and landing ships today than in the 1980s. "The construction of a new 11,000 ton Dayun class AK that carries two Super Frelon SA-321 helicopters for vertical replenishment is seen as an important addition." In overall terms, today, the PLAN is better prepared to conduct amphibious operations than it was in the late 1980s.

Naval Aviation

The one major deficiency in this area is the absence of an aircraft carrier. The Chinese have been looking around for possible imports or refits from Russia but perhaps to avoid raising undue alarm in the immediate neighbourhood, the PRC has not so far given an aircraft carrier to the PLAN.

The Russian Varyag was one such a candidate but not much is known about its progress. China would be looking for at least three aircraft carriers if it is serious about remedying the shortfall in the PLAN's air defence capability. In the absence of one, it has to depend on shore-based air defence aircraft and that naturally restricts the reach of the Chinese navy.

The PLAN continues to have some 18 H-6 (Tu-16) bombers on its inventory. Capable of carrying the YJ-6 and YJ-61 anti-ship missiles, this is a useful asset but its continued availability is not known due to its vintage. The old bombers are also extremely vulnerable to enemy action as their self defence capability is limited. The fighter force comprises some 250 J-6, 30 J-7, about 48 J-8/8A/8IIA/8B/8D but the recently acquired Su-30 is conspicuous by its absence. The range of the fighter force is about 400-500 km. The strike component likewise is also modest with some 30 Q-5 and 18 JH-7 with the latter getting a new ramjet powered ASM with considerable stand-off range. It is widely reported that the PLAN would soon get many more JH-7 fighter bombers and the J-10 which is now in production.

The PLAN is also developing its Maritime Reconnaissance – Anti Submarine Warfare (MR ASW) capabilities with the recent addition of the Y-8X, a modified An-12 with the Search-water radar of British origin. China's search for the Airborne Warning and Control System (AWACS) ended in a temporary failure when Israel cancelled the Phalcon deal in July 2000, but the PRC's efforts to acquire one from Russia have not slowed down. Once this force-multiplier joins service it would dramatically change the scenario.

Assessment

The PLAN is undoubtedly well on its way to becoming a modern and formidable navy in the next seven to ten years. The only deficiency it currently faces is that it does not have an aircraft carrier. It is known that the PRC leadership is keenly scouting for one but why it has not succeeded so far is a mystery. The carrier no doubt is a costly proposition but it is not for lack of funds that the decision to acquire one is delayed. I suspect that the PLA leadership is simply not getting one of its choice nor is it confident of quickly training its forces and reorganising the navy around one or two carrier task forces. Five nuclear powered submarines are one too many if China's objectives are limited to covering South China Sea in peacetime. With the development of the indigenous Song it is possible that the PLAN would focus on diesel-electric powered submarines rather than depend on the old and difficult Hans. The acquisition of a carrier and building its aviation capabilities are two decisions that would fundamentally change the navy's profile from an inward looking coastal defence force to a full blooded blue-water navy. The PLAN has been active in carrying out goodwill visits to countries of the world, including New Zealand and the UK and smaller neighbours and friends in the Asian region. This would give it the necessary confidence to operate away from the country's shores and by the end of the current decade, the PLAN flag is likely to be flying in many new seas and oceans of the world. David Shambaugh, a respected analyst, believes that, "...a fleet of 50 surface combatants (of which only three destroyers and four frigates approach modern standards) and a few SSN attack submarines do not constitute a blue-water navy or a power projection capability. It will require about 20 Luhu destroyers and an equal number of Jiangwei frigates to approach this level."[14] But he said this in 1997. Since then, the PLAN has received two Sovremenny destroyers and increased its inventory of the Jinagwei to 12 and added two more Song and Kilo submarines each. It is, thus, possible and, in fact, likely for the PLAN to achieve a modest force projection capability by 2010.

Notes

1 David Shambaugh, in David Shambaugh and Richard H Yang (ed), *China's Military in Transition* (Oxford: Clarendon Press, 1997), p. 25.

2 PRC Defence White Paper- December 2002, Internet version

3 David Shambaugh, n.1, p. 24

4 Ibid, p. 25.

5 Ibid.

6 John Wilson Lewis and Xue Litai, *China's Strategic Sea Power* (Stanford, US: Stanford University Press, 1994), p.4. This in fact is a fascinating account of the trials and tribulations of the PRC's strategic submarine and missile programme. The authors also throw much light on the PRC's national maritime strategy and the evolving role of the PLA Navy.

7 Ibid, pp. 4 and 215.

8 Ibid, p. 216.

9 Shambaugh, n.1, p. 214

10 Yihong Zhang, "China's Rising Forces", *Janes International Defence Review* (August 2002), pp. 36-39.

11 Ibid, p. 214 and *IISS Military Balance 2002-03*, p. 146.

12 Zhang, n.10, p. 39.

13 Ibid, p. 211.

14 Ibid, p. 28.

6

PLAAF: The Sword Arm of the PRC

Introduction

The PLAAF needs our special attention as it is likely to become the sword arm of the PRC's power projection capabilities. The PRC military leadership is fully aware of the role of air power in recent conflicts and, thus, naturally lays great emphasis on sharpening its offensive capabilities. A mere bean count is, however, not adequate to understand the real strength and, indeed, the weakness of Chinese air power. After languishing as an almost obsolescent force for over three decades, the PLAAF has, since the early 1990s, maintained a vigorous pace of progress. The decision to acquire large numbers of Su-27s and later the Su-30s, from Russia was the most noteworthy event in its march towards modernisation. It must, however, be made clear that mere acquisition does not translate into capability and in turn capability need not necessarily turn into usable power. Even so, we would do well to watch the spectacular growth of the air arm of the PLA.

For many years in the decades of the 1960s, 70s and 80s, there seemed to be no visible progress or modernisation of the Chinese armed forces. What with the high level of secrecy that the PRC leadership maintained, even the *IISS Military Balance*, a respected strategic journal, and *Jane's All the*

World's Aircraft continued to repeat the same figures about the PLAAF strength for many years. Even these were often based on conjecture. The PLAAF, it seemed possessed a huge number of F-6 (Soviet MiG-19) and a whole host of other Russian origin aircraft, which had become obsolete but could not be ignored because of their considerable numbers. The PLAAF was hardly used during the long Vietnam War. Only some Air Defence elements were active in North Vietnam for some years. Again, during the 1979 invasion of Vietnam, the PLAAF was mobilised but not used.

Even after the arrival of Deng Xiaoping on the national scene and despite the fact that he fully understood the role of modern air power, the PLAAF received little funding for real modernisation. As noted earlier, the subordinate position of the PLAAF commander vis-à-vis the PLA top brass did not help matters. One other factor that inhibited the overall rejuvenation of the PLAAF was the interminable debate over self- reliance and the absence of technology. Try as it might, China could not simply produce modern weapon systems without importing high technology.

In light of this, the decision to induct the fourth generation Su-27 in 1992 was momentous. The PLAAF has not looked back since, although there remain many weak areas, which in all probability, the PRC would address and remedy by the end of the current decade. Once that happens, the world in general and the Asia-Pacific region in particular will not be able to dismiss China's military capabilities.

Organisation

The PLAAF is organised along the same regional lines as the PLA with a Regional or Military Area Command HQ alongside the HQ of the army elements but not necessarily collocated with it. The PLAAF commander is based in Beijing with a rank and status equal to a military regional commander. On many occasions in the past, the commander was not a pilot nor did he have any specific experience in the employment of air power. Some China scholars contest this fact and repeat that all PLAAF commanders have been veterans of the Korean War. At least one PLAAF commander was purged for his close links with Lin Biao, who was suspected for the failed coup attempt against Mao in 1970. In fact, the PLAAF came under a cloud after this incident as its members were suspected of being loyal to Lin.

The air elements comprise, Air Defence (AD), fighter, bomber, attacker, short-range missile, transport, helicopter and other combat support and logistic units. Of late, much has been done to improve the logistic support function of the PLAAF with the avowed aim of expediting the supply of spares for a variety of aircraft to bases away from their peacetime locations. This shows an interest in improving aircraft availability and sortie utilisation rates. This may also indicate that in the very near future, the PLAAF may be able to deploy quickly to different theatres. The air bases sometimes have more than one type of aircraft, i.e. they may have transport, helicopter, and fighter assets at the same base. There is, however, little information in open sources on the actual deployment of the PLAAF except some reports that the Su-27 fleets were first based in Wuhu (Anhui Province), with some aircraft in Beijing MR and at a base near Kunming in Chengdu Province. The aircraft assets are organised in air divisions, regiments, and squadrons, but their strength and number per unit are not standard; the bomber divisions possessing some 70 to 90 aircraft in a division while a fighter division may have as many as 70 to 124. "A typical air defence regiment has three squadrons of three flights; each flight has three or four aircraft."[1] Similarly, it is not known if the AD missile units have a fixed number of launchers and/or missiles and anti-aircraft artillery. The PLAAF AD assets are, however, considerable; with about 100 SAM sites and over 16,000 anti-aircraft guns. It is also reported that the obsolete MiG-19 or J-6 fighters are stored in large numbers in central China with the stocks being used for keeping smaller numbers of the type in service, through 'cannibalisation'.

In the recent past, the PLAAF has inducted some air refuelling tankers and advanced AD missiles such as the S-300 and TOR-1M. Their deployment is not known but in all likelihood, these high-tech weapon systems may be based in the near vicinity of Taiwan in Nanjing MR for obvious reasons. The navy's air arm is entirely shore based but their actual locations are not known. These are likely to be evenly distributed along the entire coastline.

Strength

With its HQ at Beijing, the PLAAF is organised in five corps roughly equal to PLA group armies. Their deployment is No. 1 Corps at Changchung, No. 7 at Nanning, No. 8 at Fuzhou, No. 9 at Urumqi, and No.10 at Datong. There are 32 air divisions including 22 fighter, 3 bomber, 5 attack and 2 transport divisions. As per the latest figures in the *IISS Military Balance*, the PLAAF has some 1,900 combat aircraft. Interestingly, this figure was 2700 in 2002-03. There has been a sizeable reduction in the J-6 (MiG19) assets. The strength of the Sukhoi fleet is fast increasing and so is the strength of indigenously built J-8 II and the J-7 variants. The strength of the H-6 (Tu-16) bomber fleet has remained at around 120 for over a decade and these aircraft are possibly on their last legs. Some of these, however, have been modified for air refuelling duties and that shows the ability of the Chinese engineers to keep at least a sizeable portion of these in operational trim for the current decade. The bombers are also likely to remain useful for nuclear delivery but their numbers would continue to dwindle. The JH-7 (also called FB-7) has been in production and is now the mainstay of PLAN strike force although there are conflicting reports about the air force refusing these and only the PLAN flying them.

The PLAAF fighter force comprises some 1,000 aircraft distributed in 39 regiments. These include some 675 J-7 (MiG-21) variants, 185 J-8 variants and 100 Su-27/UBK and J-11; the last being the locally produced version of the Russian Su-27. The FGA or attack element is made up of 350 J-6 (MiG-19), 300 Q-5C and 58 Su-30 MKK. The PLAAF is reported to possess some 290 reconnaissance and Elint aircraft including some old types like the HZ -5, HZ-6 and JZ-& and 4 Tu-154 aircraft. The J-10 and FC-1 are already flight tested and now with the Russian engines for both types, their series production is assured in the next year or two. The PLAAF transport fleet has some 513 aircraft including 15 Tu-154, 350 Y-5 and Y-7 and some 50Y-8 (An-12).The Chinese are building the Y-8, a local but modified version of the AN-12 with new engines with more efficient propellers. The PRC is also already producing a whole host of helicopters and AN-26 class of tactical transports and it is reasonable to assume that with the Western aircraft manufacturers forecasting China as the biggest commercial aircraft market in the coming years, the PLA would not be short of strategic heavy lift resources.

The PLAAF relies on some 200 trainers of all types for its training requirements but this figure does not appear to be very accurate as 200 trainers are obviously inadequate for an air force of 2000 combat aircraft. The information about its AD assets is also sketchy with the HQ-2, 2A, 2B, versions of the old SAM-2 missiles continuing to feature in this category.[2]

Research and Development

As already noted in an earlier section, the PRC's long-term goal is to acquire expertise to build its own high-tech weapons, equipment, and aircraft. Some notable successes are already apparent but the PRC need not worry so long as it can obtain aircraft and AD missiles from Russia. Here we shall discuss only air power related developments.

The PRC defence industry started producing the J-7, a local version of the famed MiG-21 in the late 1960s. There were reports that the Chinese had illegally impounded some MiG-21 aircraft that were on their way from the Soviet Union to North Vietnam, and used them to produce their own through reverse engineering. Notable successes in this field are:

The Chinese engineers increased the area of the wing of the MiG-21 by about 15 per cent and added leading edge flaps thereby increasing the overall lift. This also improved its manoeuvrability in combat and should have reduced the rate of descent on the final approach and the landing speed.

The original R-11/13 Russian engines were manufactured in the country and their thrust increased by about 15 per cent. This would have improved the total all up weight of the fighter.

Other modifications included the placing of armament switches in more convenient locations, additional under-wing pylons for carriage of armament and or drop (fuel) tanks, improved avionics and the ability to carry Western armament such as the French Matra Magic air-to-air missile.

The more recent improvements include a double delta wing plan form changing the very look of the aircraft.

The PRC has also used refurbished Rolls Royce Spey engines (originally fitted on the Phantom F-4) for the indigenously produced JH-7, again a derivative of the MiG-19, although there is little similarity between the two aircraft.

Both these fighters have been modified for air-to-air refuelling which must increase their range considerably.

The Soviet AN-12 has also been extensively modified with new engines (Pratt & Whitney Canada PW-150A) and

Doughty composite eight-bladed swept propellers, thus, improving its payload capacity and overall efficiency. Fitted with the 'British Searchwater' radar, the Y-8, as it is called in China, is the primary aircraft for maritime patrol duties. It is also reported that the new aircraft would cost much less than the US C-130, and hence may soon prove an attractive export option.[3]

The PRC also produces a variety of French helicopters under licence and has also designed a few of its own.

China has been pursuing cruise-missile technology for some time and claims to produce a number of air, surface, ship, and submarine launched cruise missiles with ranges from 250 to 1800 km. We have already noted the continuing fast-paced production of DF-15 (M-9) and M-11 SRBM with conventional warheads. These missiles were fired across the Taiwan Strait in 1995 and 1996 when the PRC successfully intimidated the Taiwanese leadership from going overtly independent.

The PRC has achieved some impressive milestones in the field of satellite launch technology; and since most of these are dual use, civil/commercial and military, these have enhanced China's communications capabilities. There is no doubt that with its own navigation, reconnaissance, and surveillance satellites, the PLA's Command, Control, Communication, Computer, Intelligence, Surveillance and Reconnaissance (C4ISR) system is fast becoming modern and secure. Having loaned some Russian GLONAS satellites, China is also reportedly planning to deploy its own Global Positioning System (GPS) satellites. If the success of the 15 October 2003 manned space flight is any indication, placing a few geo-synchronous GPS satellites in space would be child's play for the PRC.

In the field of electronic warfare, information security, encryption software, high-speed data transfer and laser technology, too, the PRC's progress has been impressive, thus, qualitatively improving the C4ISR capabilities of the armed forces.

Another important pursuit for local R&D is the production of a variety of AD, air-to-surface, air-to-ground missiles and ram jet powered stand off precision weapons. The PRC has already displayed some of these at the annual Zhuhai air shows.

New Aircraft Design and Development

The PRC's record of aircraft design and development was dismal in the 1960s and 1970s. It was reported that at one time the Chinese had as many as 23 new designs on the drawing board with little success.[4] The quality of the indigenously produced A-5 (also known as F-5 and Q-5) and the J-6 was so poor that a huge number of these had to be returned to the manufacturers for extensive repairs; even airframe parts would not fit the aircraft. Since then, the PRC's defence industry has come a long way with at least two fighters, the J-10 and FC-1 ready to roll out in the very near future. The former is reportedly based on the Israeli Lavi and is similar to the US F-16 in appearance and performance. Pakistan had allegedly provided the Chinese with a crashed F-16 airframe for design work. This aircraft was languishing for many years on the drawing board for want of a suitable engine. The 1999 deal with the Russians for the supply of 300 Al-31 F Saturn Lyulka turbofan engines that are fitted on the Su-27/30 was a 'coup' of sorts for the PRC. It can now produce the J-10 in large numbers with some 54 engines already delivered to the factories.

The continuation of the FC-1 programme was also in serious doubt for two different reasons. First, the PLAAF was not ready to commit funds for it while Pakistan, the programme partner, did not have the necessary funds. Second, this fighter too did not have a suitable engine until the Russians came to the Chinese rescue and offered to sell them large numbers of RD-

93 engines. This engine is a variant of the RD-23 that has been successfully used on the MiG-29 air superiority fighter for many years. With this engine in their pocket, the Chinese are set to produce the FC-1 in large numbers and Pakistan would reap the benefits of perseverance and trust in its old ally.

The PRC reportedly plans to produce the XXJ stealth fighter, but not much information is available in open sources. Given its recent record of successes, it is quite likely that the PRC would soon produce this fifth generation fighter too.

Training, Doctrine and Operational Preparedness

Much has been written on the PLAAF's doctrine and the newly acquired hardware; there is however, little analysis of its true or actual fighting capability. As noted before, the PRC leadership did not use its air power assets except during the 1950 Tibet invasion, the Korean War, and its air defence elements in the many crises and skirmishes in the Taiwan Straits. During the Korean War, its performance was far from satisfactory but that can perhaps be attributed to inexperience of the young pilots and the general superiority of the US fighter aircraft. The amazing fact, however, is that the PLAAF has hardly seen action since then, although China has been involved in armed conflicts at regular intervals until the invasion of Vietnam in 1979. In the 1995-96 Taiwan Straits exercises, the PLAAF fired only a few missiles and hence its participation was, at best, low key. Yet in the recent past, it has acquired a wide array of weapon systems and its commanders regularly talk of the air element being in the vanguard of any future conflict. What then is the ground reality? Is the PLAAF really capable of wielding the fourth generation Su-30 and derive the most from this multirole fighter? In the absence of reports about exercises, there are no easy answers.

Most China watchers agree that over the last decade, PLAAF exercises have become increasingly more realistic in form and content with jointmanship being given greater importance. Many Chinese language reports also speak of exercises being conducted with greater realism. More and more PLAAF squadrons are now capable of flying by night, over water, in bad weather and conducting live firing exercises. In other air forces, including the Indian Air Force, fighter squadrons routinely carry out these exercises and all squadrons regularly train new entrants until they reach a minimum standard of proficiency and the more experienced aircrews practise air combat, live firing and dissimilar aircraft combat training as part of annual routine and, hence, is not highlighted in the same manner. Curiously enough, most Western writers focus on larger strategic issues and inexplicably ignore the 'real content' of flying training. Merely referring to annual flying per pilot does not indicate the skill levels achieved. For example, in one of my earlier papers, I had commented on a report highlighting the number of 'successful' arrester barrier engagements. For the uninitiated, an arrester barrier is installed at the end of the runway to prevent a high-speed fighter or trainer from overshooting the paved area and is used only in an emergency. The more the engagements the more the accidents; these are not an indication of efficiency.[5]

The PLAAF has for some years, been exercising with a 'blue' aggressor squadron. This appears to be the direct result of close interaction with the US Air Force during the heyday of Sino-US military-to-military cooperation, but since these exercises are also shrouded in secrecy it is difficult to assess their impact on the operational preparedness of the PLAAF. Similarly, in my earlier paper, I had questioned the statistics on the actual flying training per pilot per year since the number of aircrews was obviously large but the availability of aircraft, especially the J-6, which was the mainstay of the fighter fleet, low. This anomaly has been resolved with the 2002-03 figures[6] showing only 1,500 J-6 as against 2,500-3,000 in the earlier years. These figures have been further revised downwards indicating a rapid phasing out of this old workhorse. The PLAAF is obviously concentrating on some Class 'A' units while the others get by with less flying. With the rapidly increasing availability of J-7 and J-8, the situation appears to be fast improving. Even so, it will take the PLAAF the better part of five years to catch up with the backlog of operational flying training and operationalise the new aircraft like the Su-27/30. It is quite probable that until the PLAAF actually achieves the desired levels of proficiency, the PRC would not publicise the readiness state of its air arm. China does not need to carry out 'fire power demonstrations' to deter a future adversary. In sum, the PLAAF is still maturing; how long that process takes is difficult to assess at present. One could surmise that it would perhaps follow the Russian rather than the Western model of training and may take five to seven years to reach acceptable levels of fighting skills and expertise. Even its commanders would need time to be able to employ modern air power effectively and independently.

Assessment

In all likelihood, the new PRC leadership comprising entirely of technocrats also understands the critical role of air power in the wars of the 21st century. China is making a concerted effort to modernise its air force. It has already acquired a number of sophisticated aircraft, radars, missiles, and support equipment. It would, however, take the PLAAF some five to seven years to absorb these weapon systems and employ them effectively in future wars. The PLAAF commanders too are fast becoming alive to the idea of 'modern' air power. They would, however, need much more freedom to hone their skills to employ air power as a political instrument for coercion and intimidation. In my opinion, this would be achieved before the end of the current decade. Russia's role as a provider of expertise, technology, and training would be crucial.

Indicators of Further Developments

Considering the important role that the PLAAF is bound to play in any future conflict, China watchers in India, need to monitor developments in the following areas. These may give rough indications of the direction in which the PLAAF might progress. A caveat is, however, necessary. These will not give any insights into the circumstances and the occasions that may encourage the PRC to use military force but may prove to be a useful guide.

- The actual basing of various types of aircraft.
- The type and frequency of air force exercises.
- The role and responsibilities of the PLAAF commander.
- Details of flying training such as flying hours, utilisation rates of aircraft, live firing exercises and the missions in training with the 'blue' aggressor squadron.
- Annual budget for the PLAAF.
- Visits of PLAAF senior commanders to Western countries and to Russia.
- The success of the J-10 and FC-1 programmes.
- The actual deployment of the air-to-air refuelling tankers.
- Peacetime deployment of the AWACS aircraft.
- Radio procedures and acquisition of secure multi channel/frequency radio sets.
- Utilisation of the UAV and cruise missiles.
- The philosophy of AD missile deployment.
- Command & Control procedures.
- The experience and success of joint commands in various military regions.

- Joint exercises with the army and navy and offshore air defence procedures.
- The search for the aircraft carrier and its deployment.
- Emergency/contingency training with China's commercial aviation assets for mobilisation.
- A regular check of the display and demonstration of various aircraft, radar, missiles, and equipment at the annual air shows at Zhuhai and elsewhere.

Notes

1 Federation of American Scientists, "PLAAF Order of Battle", available on the internet at http://www.fas.org/nuke/guide/china/agency/plaaf-orbat.htm

2 *IISS Military Balance 2003-04*, (London: International Institute of Strategic Studies, October 2003), pp. 154-155.

3 Richard Fisher, 'The PRC's third Zhuhai Air Show- Nov 5-12,200', lecture delivered at the Center for Security Policy, 7 December 2000 and personal discussions with the author at Delhi and Washington DC.

4 John Wilson Lewis & Xue Litai, "China's Search for a Modern Air Force", *International Security*, Vol. 24 (Spring 1999), p. 89.

5 See for example, RV Phadke, "PLAAF- Shifting Air Power Balance: Challenges to India's Security", CISAC Working Paper, February 2002, available on the internet at http://www.cisac.org/htm or http://www.stanford.edu

6 *IISS Military Balance 2002-03* (London: International Institute of Strategic Studies, October 2003), pp. 173-178.

7

Nuclear and Strategic Assets

Introduction

China and India share a number of important attributes. Exceptionally large population, huge continental size, old civilisation, culture and a history of invasions, internal turmoil and foreign exploitation are some of the major similarities. Even in their post-independent history, both suffered from widespread poverty, dependence on agriculture, natural calamities, social and financial inequalities and a myriad other impediments to national development. Their approaches to national security, however, were completely different. China chose to send its armies to 'liberate' Tibet, and entered the Korean War in support of North Korea, which was clearly the aggressor; India chose to go to the United Nations when it was attacked by Pakistan in J&K. Nowhere else is the difference more stark than in their approach to the issue of nuclear weapons. While India felt that possession of nuclear weapons was immoral and rooted for total nuclear disarmament, the PRC chose to build one of its own even though its ally, the Soviet Union, had extended the nuclear umbrella over China. The US had threatened China with a nuclear attack during the Korean War and Mao did not like to be intimidated. He also did not accept the idea of only the

Soviet Union possessing nuclear capability. "Chinese leaders asserted that the Soviet Union's status as a nuclear superpower should not be made a justification for preventing other socialist countries from increasing their own defensive capabilities".[1] Mao realised that the PRC could not depend on another country for nuclear protection. He knew that possession of nuclear weapons was an urgent necessity if China were to count as a big power. The saga of China's efforts to acquire nuclear weapons is indeed a shining example of successful manipulation of allies and adversaries. Mao, and later his worthy followers cleverly 'employed' the PRC's nuclear assets to ward off national security threats when the country's conventional forces were completely inadequate.

Early Years

Accordingly on 15 January 1955, a nuclear weapons programme codenamed 02 was launched in a meeting presided over by Chairman Mao. Being a close ally, Mao naturally turned to the USSR for assistance which came in the form of raw material, technology, specialists, technicians and training, but when it came to giving a sample of the bomb the Soviets declined, although the two countries had signed a treaty for cooperation in the nuclear field. Mao saw this as a betrayal by the Soviets and this was to become the major cause of the rift between the two.

Work nevertheless went on; and in a surprisingly short period, the Chinese tested an atomic bomb–a conventional fission device–on 16 October 1964. While it is known that the former Soviet Union assisted the PRC, the sheer scale of the assistance is overwhelming and, therefore, needs further elaboration here. The Thirty Year Treaty of Friendship, Alliance and Mutual Assistance signed between the PRC and USSR on 14 February 1950 set the stage for the flow of massive military and scientific-technological assistance. The doors were now open to hundreds of Chinese scientists to train in the USSR and for thousands of Soviet scientists and technicians to work in China to build what became one of the largest foreign funded and assisted defence industry programmes. "Nuclear cooperation began with the establishment of a Sino-Soviet Non-Ferrous and Raw Materials Company; this was the beginning of exploration and uranium mining in Xinjiang".

"Between 1955 and 1958, six accords were signed relating to Soviet assistance in the development of China's science, industry, and weapons programmes. These included joint surveys of Chinese uranium ores, supply of a nuclear reactor and a cyclotron, and in building China's nuclear research facilities and industries. The New Defence Technical Accord of October 15, 1957, however, was unusual in the history of the nuclear age; the Soviet Union promised to supply China with the blueprint for and a working prototype of an atomic bomb and missiles as well as related data".[2]

Mao saw that the only way to expedite the process of bomb making was to get a sample from the Soviet Union. China had to extract the maximum help from the Soviets. Following the uprisings in Poland and Hungary, Soviet behaviour and actions were not exactly liked in those countries. The Chinese leaders played an important role in restoring Soviet influence in Eastern Europe. The Chinese also knew that Khrushchev would need their support in the power struggle then underway at the Kremlin. The Chinese extracted their pound of flesh by making this apparently extraordinary demand for the prototype of the bomb, confident that they would get it. This is another example of the Chinese art of exploiting all possible opportunities. Negotiations for a Partial Test Ban Treaty, however, came in the way. Khrushchev did not want the transfer of the bomb design to China to spoil the chances of building a peaceful relationship with the US and refused to provide the sample of the bomb. The Communist Party of the Soviet Union (CPSU) letter declining the Chinese request for the prototype of the bomb was dated 20 June, 1959. So furious were the Chinese that they codenamed the first device '596' commemorating the year and month of the Soviet refusal[3]. So much for the Chinese penchant for symbolism.

Under the New Defence Technical Accord of 15 October, 1957, Khrushchev had earlier sent an army missile battalion with two R-2 missiles (called the SS-2 in the West), with a range of 590 km and their associated launching equipment, blueprints and documents in 10,151 volumes and missiles were also sent to China for manufacturing, testing and launching purposes. "The PRC was, however, unhappy with the Soviet offer of R-2 missiles, as these could not carry the Chinese warhead. The Chinese were looking for the R-12 (SS-4) the then state-of-the-art missile, which could. The Russians refused to give them the R-12. Undeterred, Chinese students studying at the Aviation Institute at Moscow acquired a rudimentary knowledge of the missile, which later provided the basis for the Dong Feng (East Wind) DF-1 missile. The PRC was once again fortunate. Although Sino-Soviet relations were already worsening, in December 1959, Khrushchev authorised the sale of the R-11 missile as well as the Golf submarine on which it could be deployed. It was the only Soviet missile capable of hitting targets in the United States." This must have come as a huge windfall to the Chinese scientists.

Some 11,000-50,000 Russian specialists helped China develop its science and technology and some 38,000 Chinese scientists received training in the USSR, including at the Joint Institute of Nuclear Research at Dubna. By the time of the Russian departure in 1960, China had a 100,000 scientists and 300,000 engineers. During the 1950s, China received massive quantities of arms and equipment from the USSR. The Soviets also helped build 24 defence enterprises and a further 33 were built with Soviet technical assistance as late as in 1963.

The success of the first atomic test was a great morale booster for the Chinese nation. Not only did it deliver the nation from the threat of nuclear blackmail, it added to China's stature. Such was the mystique of China's first nuclear explosion that although many major powers condemned it, in fact the number of countries recognising the PRC regime increased following the tests. Soon thereafter, the PRC successfully fired a missile and exploded a hydrogen bomb on 17 June 1967. China had developed the thermonuclear bomb in a remarkably short time.

The Americans knew of the prospect of the PRC testing its nuclear device. They had, for a while toyed with the idea of striking at the Chinese nuclear facilities at Lop Nor but soon abandoned the plans. This was to be a joint US-Soviet strike with one bomber each from the two countries dropping one live and one safe bomb,[4] a strange arrangement between two nuclear adversaries to keep the third out of the race.

There was also the Senator McGhee plan to offer India nuclear assistance to build its own deterrent as a counterpoise to the Chinese bomb but once again, the idea never bore fruit because the Americans thought that Nehru would not opt for nuclear weapons and join the Western bloc against the Sino-Soviet combined threat.[5]

The Chinese, however, soon declared their intention never to use the atom bomb before being attacked by the enemy, adding the NFU or 'No First Use' doctrine to the nuclear lexicon. The PRC leadership stated that China stood for total disarmament and that they were aiming to develop only a "minimum deterrent capability", which they would never use unless first attacked, a position that has remained largely unchanged for the last forty years. The NFU pledge saw some subtle changes when in the early 1990s the PRC decided to support the international non-proliferation regime when it became clear that staying outside the regime would further isolate China and not serve its overall national interests. China's objective of building a credible deterrent was neither diluted nor abandoned. During the 1969 Usuri River clashes with the former Soviet Union, the PRC did not allow the situation to escalate but was obviously not ready to bend to Soviet dictates; a successful if subtle political use of nuclear weapons in a crisis. Even during the ten-year-long Cultural Revolution, which targeted scientists and intellectuals and forced the closure of most colleges and universities, work on nuclear weapons continued without any disruptions. In fact, it received a boost with the PRC's decision to continue the work on the nuclear submarine. The PRC wanted a survivable deterrent. Work on Project 09 for the construction of the nuclear submarine had begun as early as in 1958 and went on with some temporary disruptions until in 1982 a submarine launched missile was tested. Soviet assistance did not stop completely even after relations between the two had begun to cool.

Foreign Assistance

Over the years, the Chinese nuclear arsenal has slowly developed into a formidable force although its numbers have remained relatively small. It is evident that the PRC leadership clearly understood the political power of these weapons and opted out of a costly nuclear weapons race of the type that finally crippled the Soviet economy. Mao was satisfied with a small arsenal just adequate for a counter attack. He and his cohorts did not get involved in 'nuclear accountancy'. This clearly shows that the PRC leadership was never interested in competing with other nuclear powers. By relying on a small but credible deterrent, China not only avoided wasting scarce resources but also maintained adequate flexibility to open negotiations with either of the superpowers, depending on the circumstances. In essence, while maintaining a limited capability to strike at targets in the continental United States, and the Soviet Union, and the US assets in the Pacific, China did not threaten either of these powers with an imminent nuclear attack. This was a clever move as proved by the events of the early 1970s.

Following the Pakistan aided, secret visit of Henry Kissinger in July 1971, China broke its isolation and entered a quasi alliance with the US against the USSR, a relationship that proved extremely beneficial to the PRC. China soon occupied the Security Council seat when it joined the United Nations in 1971, obviously as a *quid pro quo* for extending the hand of friendship to the US. This not only gave the PRC the veto power but also raised its international prestige now that it was a member of the Five Power Nuclear Club. In the bargain, this legitimised its status as a major Asian power and afforded it the opportunity to sup at the high table. This is significant considering that at the time the PRC was not only facing international isolation but was a poor developing country experiencing the deep turmoil of the pernicious Cultural Revolution. This is yet another example of China's clever use of its nuclear capability.

In the early 1970s, there came another windfall, this time from America. With the improvement of relations with the US, "successive US administrations viewed a militarily strong China as a welcome counter balance to the USSR. Even without any Chinese request, senior American officials in the Nixon Administration supplied sensitive intelligence data on the Soviet military capabilities and deployments. Henry Kissinger facilitated the sale of two large computers to China in 1976. The Ford Administration encouraged major European powers to sell weapons to China and provide access to Western technology. Military ties were deepened during the Carter presidency and China was allowed access to dual-use American technology. With the beginning of the crusade against the Soviet Union's Afghan adventure, the Reagan Administration decided to accelerate technology transfers to China. Technology controls were relaxed and China imported avionics packages, anti-submarine warfare torpedoes, advanced computers, and sophisticated machine tools. American satellites were launched from Chinese launch vehicles and rockets, known to be part of, or derived from, the Chinese nuclear weapons programme. Exports of military equipment were banned after the 6 June, 1989 Tiananmen incidents but the sale of dual-use technology, including lasers, computers, and test instruments continued. Scientific exchanges continued with American scientists from the Los Alamos and Livermore laboratories visiting Chinese nuclear establishments and test sites at Lop Nor. The '863' programme proposed by Deng Xiaoping and designed for research in cutting-edge technologies also received a fillip with these exchanges.[6]

William Perry, an early advocate of 'smart and stealth' weapons, also held talks with senior Chinese generals, giving them insights into the area of American fourth generation technologies. Apart from these high level contacts with the American defence establishment, China has also been directly importing technology through leading American aerospace companies eager to tap the vast China market. "The Clinton administration effectively removed all controls over the sale of a whole category of dual-use technologies. American companies could export to China computers seven times more powerful than those sold in 1976. China promptly purchased some one hundred supercomputers that were ideally suited to perform the vast number of calculations needed to simulate nuclear explosions."[7]

Strength

The capabilities of the Chinese nuclear aircraft and missiles are given below. China defines missile ranges as follows: short-range, <1,000km, medium-range, 1,000-3,000km; long-range, 3,000-8,000 km; and intercontinental range,> 8,000 km.

Table 4.1

Type	NATO Designation	No. Deployed	Year first deployed	Range (km)	Warhead yield	Warhead in stockpile
Aircraft H-6	B-6	120	1965	3,100	1-3	120
Q-5	A-5	30	1970	400	1 bomb	30*
Land-based missiles						
DF-3A	CSS-2	40	1971	2,800	1x 3.3 Mt	40
DF-4	CSS-3	20	1980	5,500	1x3.3 Mt	20
DF-5A	CSS-4	20	1981	13,000	1x4-5Mt	20
DF-21A	CSS-5	48	1985-1986	1,800	1x200-300 Kt	48
SLBM						
Julang I	CSS-N-3	12	1986	1,700	1x200-300 Kt	12
Non-strategic weapons						
Artillery/ ADMs Short-range Missiles					Low Kt	120
Total						-410

* Note: These attack aircraft are in the process of being phased out and by the end of the decade would be replaced with newer types.[8]

In addition to the missiles listed above, the DF-31 has long been considered to be in the final stages of development. A 1999 report that the DF-31 has been deployed in southern China has not been confirmed. The DF-31 is a three-stage solid propellant, mobile Inter Continental Ballistic Missile (ICBM) with a range of some 8,000 km and has a Circular Error Probable (CEP) of 0.3-0.5 km. This missile has been flight tested three times, the latest on 4 November 2000, with several decoy warheads. A report states that in 1987, the missile underwent trials that simulated a launch from a submarine. A variant of the missile is believed to be under development to arm the new Type 094 Class SSBN. There were media reports of a simulated test of the DF-41 in 1999, but this missile is thought to have been cancelled.

According to the Stockholm Peace Research Institute (SIPRI), the nuclear capability of the 600-km range M-9 (CSS-6) and the 300-km range M-11 (CSS-7) is unconfirmed. The Taiwanese Defence Ministry has specifically referred to the M-9 and M-11 as nuclear capable. One regiment size M-9 unit is deployed in southeastern China and may soon be augmented by an additional unit. Reports of China building up to 650 M-9 missiles have been widespread in the media and US intelligence assessments. According to a US report, the PRC already possesses some 450 Short Range Ballistic Missiles (SRBMs) deployed specifically for a possible conflict against Taiwan.[9]

Allegations of Chinese theft of US nuclear warhead designs have fuelled speculation that China may soon deploy missile systems with multiple warheads. China has had the technical capability to develop MRV payloads for over 20 years A MRV capability should not pose a major difficulty if the DF-31 is used as the launch vehicle.

Nuclear Bomber Force

China has maintained a modest force of about 120 H-6 (Chinese version of the venerable Soviet designed Tu-16) for nearly four decades. These aircraft are by all accounts ancient and yet the PRC has modified some of these for aerial refuelling operations. This means that their reliability and maintainability is still acceptable. Their ability to deliver nuclear bombs after negotiating the air defences of the adversary is, however, quite doubtful and hence it is likely that China would phase out these bombers by the end of the current decade. To retain the necessary flexibility, the PRC may choose to employ the Su-30 in the nuclear role. "China purchased 40 Su-30 multi-role aircraft from Russia in 1999 at a price of $2 billion" and is likely to build some 200 Su-27/30 combat aircraft under licence. It is also well on the way to produce the J-10 and the FC-1. This would give China considerable flexibility to choose a more suitable replacement for the antiquated H-6 (Tu-16) bombers.[10]

Nuclear Artillery

There are reports that China possesses a small number of tactical nuclear weapons in the form of artillery but there is little supporting evidence available. Some claim that with the change from a doctrine of minimum deterrence to one of limited deterrence (discussed later), China may decide to undertake a major upgrade of its weapons arsenal and build tactical nuclear weapons, but so far the supporting evidence is very meagre.

Rumours that China has deployed nuclear weapons in Tibet have not been confirmed. The Indian Ministry of Defence (MOD) Annual Report states that China's missiles' cover all the important Indian cities but naturally does not speak of the missiles locations.[11] One school believes that given the range and numbers in the Chinese missile arsenal there is no real need for the missiles to be deployed in Tibet where problems of storage, maintenance and even reliability may be acute due to the terrain, high altitudes, low temperatures and poor logistics. One Chinese scholar, in a private conversation with the author, ridiculed the idea of deploying missiles in Tibet for the same reasons but chose not to comment when it was pointed out that the ability of a country to widely disperse its small nuclear arsenal would add to deterrence making a counter force strike more difficult. It is, thus, safe to assume that with the completion of the Golmo-Lhasa railway line, China is bound to move some rail-mobile missiles to the Tibet Autonomous Region (TAR). Other infrastructure developments in the region would also make the operation of these missiles easier in the near future, with more and more aircraft, radars and other military units regularly operating from the desolate region of Tibet.

One difficulty, nonetheless, would remain. The Chinese armed forces are recruited, operated and maintained more or less on a regional basis although inter-regional transfers of senior officers are not uncommon. This makes it relatively difficult to find enough numbers of suitably trained and motivated personnel, especially technicians, for operations in distant TAR. It is, therefore, quite likely that whenever China decides to deploy its nuclear weapons in TAR, their numbers may remain small, especially since those stationed in Sichuan, Yunnan and Chengdu would adequately cover most of the Indian subcontinent.

Limited Deterrence

Alistair Iain Johnson, an American specialist on China's nuclear strategy, says, quoting a number of Chinese scholars, that in recent times some Chinese analysts have begun to doubt the efficacy of the concept of minimum deterrence.[12] They seem to think that to really deter a potential enemy, China must in fact think of how it would use its arsenals in an actual war. Without such definitive articulation of their employment, a second strike by a few weapons would be ineffective as the adversary would still have the means to prevail. They, therefore, advocate the development of a variety of more accurate and long-range weapons of differing yields to afford the necessary flexibility to face future nuclear threats.

This would entail a major overhaul of Chinese missiles and other tactical weapons. He contends that "at present, China, at best, has a strategic monad, with most of its operational capabilities concentrated in predominantly liquid fuelled, land-based missiles of different ranges. The one operational SSBN is in the docks most of the time. China's strategic bomber force is extremely outdated." The PRC has, no doubt, developed the M-9 and M-11 class of SRBMs but it is not clear if China has "designed a warhead that is small enough to make these missiles nuclear capable." So it is not clear if China indeed has any theatre or tactical nuclear weapons. Given the present state of the PRC's assets, it is safe to assume that China would incrementally upgrade and improve these. In all probability, China is likely to continue working on the DF-31 solid fuelled missile and use its variants for its submarine force, as and when the next generation of the nuclear submarines are ready. A dramatic change in the quality and numbers of China's strategic assets does not seem to be on the cards in the near future.

Foreign Policy and Nuclear Capability

China's move to build a large number of short-range missiles was to address the gaps in its conventional weapons capabilities for a possible future conflict with Taiwan. In the absence of modern conventional forces, the PRC needed an insurance policy to prevent Taiwan from declaring independence. A ready and deployed force of some 350-450 SRBMs in the south of China has been a major deterrent to Taiwan. These missiles were also designed to stay within the Missile Technology Control Regime (MTCR) stipulated ranges of 300 km. In 1988, the PRC sold the 2,000 km-range CSS-2 or DF-3 (with conventional warheads), to Saudi Arabia. In return, Saudi Arabia was forced to sever its diplomatic ties with Taiwan. The secondary aim was to improve China's stock among the Muslim countries of West Asia. It was also a blunt political message to the US that the PRC was not ready to abide by any international treaties if these were unfavourable or unequal in China's view. China not only gave Pakistan the much needed nuclear know how including the prototype of its 1964 test bomb but also sold some 500 ring magnets used in uranium enrichment. Further, the PRC gave Pakistan some 80 M-11 missiles which some analysts categorise as nuclear capable. For many years, Beijing resisted US pressure to stop its nuclear and missile assistance to Pakistan. Some Chinese strategists openly equating the Sino-Pak ties with the US-Israel special relationship, clearly indicating that China would never stop assisting Pakistan in enhancing the latter's military and nuclear strength.[13] Although of late, China has shown some indications abide by international regimes, it has not yet given a firm commitment. The US continues to downplay this in order to persuade China to become a more responsible member of the world community. The sale of missiles to Pakistan was in retaliation to the 1992 US decision to sell 150 F-16 fighters to Taiwan. Since 1996, China has observed a self-imposed moratorium on nuclear testing and has taken steps to join the various non-proliferation regimes and arrangements.

For many years after the 1964 test, the PRC did not show any interest in the international non-proliferation regime calling it a tool of imperialism to prevent others from acquiring nuclear capabilities. It did not join the Nuclear Non-Proliferation Treaty (NPT) or adhere to the Partial Test Ban Treaty. Many of its nuclear tests were in the atmosphere. China also did not raise very serious objections when India tested its "peaceful nuclear device" in 1974. It is significant that India's attempts to normalise relations with China started showing results only after this Indian demonstration. According to John Garver, China decided to join the nuclear non-proliferation regime at the end of the 1980s when its covert assistance to Pakistan's nuclear weapons programme had successfully resulted in the latter developing a nuclear capability. The PRC had always steered the nuclear debate so that it remained focussed on India-Pakistan nuclear rivalry. This way, India would forever remain tied to the subcontinental nuclear calculations and would be seen a relatively small player in a category of second grade powers. To serve this objective, China had always supported a nuclear free South Asia. Such a stance would have served two purposes: first, covert Pakistani nuclear capability would ensure that India could never dominate its smaller neighbour; and, second, India's nuclear ambitions would always be seen in the India-Pakistan contest thus securing a nuclear free neighbourhood for China. During the 1995-96 deliberations in the Conference on Disarmament (CD) on the NPT and the Comprehensive Test Ban Treaty (CTBT), China changed its position several times and supported the US/UK proposal to foreclose India's nuclear options by forcing it to accede to the CTBT as a non-nuclear weapons power. India, however, did not succumb to international pressure and refused to sign the CTBT that it thought was expressly designed to constrain India's options. The strident Chinese criticism of the Indian nuclear tests of May 1998, thus, came as no surprise to India. The Indian leadership also had no option but to cite the long-term Chinese nuclear threat as without doing so the world would have once again treated India on the same level as Pakistan. Having failed to deter India from going overtly nuclear, the PRC is now slowly coming around to manage a newly rising nuclear India through renewed diplomacy but it has not abandoned its support to Pakistan.[14] The PRC has, thus, very efficiently used its nuclear capabilities to achieve its foreign policy objectives. In the last years of the Clinton Administration, the American plans to +position National Missile Defence/Theatre Missile Defence (NMD/TMD) on Taiwan appeared quite plausible. It was thought that the PRC would retaliate by quickly increasing its nuclear arsenal. So far, there appears to be no indication of any hectic activity on this account although the PRC continues to upgrade its missiles and weapons at its usual pace. As in other fields, China's strategic modernisation plan has received much attention, but it is not clear if its efforts are aimed at deploying a much larger and more modern force.[15]

Conclusion

It should be clear from the foregoing that a mere bean-count of China's nuclear weapons and missiles does not illuminate the strategic picture. The PRC is no doubt modernising its conventional and nuclear forces but its real force multipliers continue to be its strategy of deception, camouflage and concealment in all its international interactions. It will take a long time for China to become transparent but even when it gives the impression of showing its hand, the real purpose is to confuse the enemy. China is definitely upgrading its nuclear arsenal at its own pace. Instead of matching the US, weapon for weapon, China may well develop a capability to defend its vital areas by improving the numerous Russian origin Air Defence (AD) missile systems such as the S-300. In the past, China has, with deft diplomacy, managed to obtain assistance from both Russia and the US. Russia once again continues to be the major benefactor for the PRC's strategic modernisation. There is little official cooperation in the nuclear field but it is a well known fact that some 2,000 Russian nuclear experts and engineers are employed in the PRC's nuclear industry.

In 1999, You Ji, a Chinese scholar at the Australian National University had said, "There have been other modernisation attempts that are transforming the SMF- Strategic Missile Forces- from a 'hiding' power to a 'fighting' power. By the year 2000, for instance, all DF-4 and DF-5, ICBMs will be placed in new generation launching silos. Most DF-3 IRBMs have been converted to solid fuel use, which will substantially reduce the preparation time for launches."[16] This has not been confirmed but the chances are that overall modernisation of China's strategic assets continues. However, these are only incremental improvements that any country is bound to make to its nuclear arsenals. With the DF-31 gradually joining the service in numbers, the Chinese arsenals are likely to remain at about the current levels. What India needs to watch carefully is the employment doctrine and actual deployment of China's SRBMs that are joining the service in very large numbers.

Notes

1 Matin Zuberi, "Soviet and American Assistance and the Pace of Chinese Nuclear Tests", *Strategic Analysis*, Vol. XXIV, No. 7 (October 2000), pp. 1247-1266.

2 Ibid, p. 1248.

3 Ibid. p. 1249.

4 Ibid. p.1250.

5 Amitabh Mattoo, " Shadow of the Dragon-Indo-US Relations and China-", in Gary K. Bertsch, Seema Gahlaut and Anupam Srivastava (eds), *Engaging India* (New York and London: Routledge, 1999), p. 219.

6 Zuberi, n.1.

7 Ibid. p. 1256.

8 Hans M. Kristensen, in SIPRI Annual Report for 2002, Appendix 6A.

9 US DOD Annual Report on the Military Power of the People's Republic of China, at http://www.defenselink.mil/pubs/20030730chinaex.pdf, pp.45.

10 Ibid. Information on China's missiles and other assets and the table are from SIPRI Annual Report.

11 "All Indian Cities Within Chinese Missile Range", Annual Report of the Indian Ministry of Defence, 30 May 2003, on Sify News at <http://www.sifynews.com.htm>.

12 Alastair Iain Johnston, "Prospects for Chinese Nuclear Force Modernisation: Limited Deterrence Versus Multilateral Arms Control", in David Shambaugh and Richard H. Yang (Eds.), *China's Military in Transition* (Oxford: Clarendon Press, 1997), pp. 290-292.

13 Andrew J Nathan and Robert S. Ross, *The Great Wall and The Empty Fortress- China's Search for Security* (New York, London: W.W. Norton & Company, 1997) pp. 151-155.

14 John W Garver, *Protracted Contest, Sino-Indian Rivalry in the Twentieth Century* (New Delhi: Oxford University Press, 2001), pp. 343-367.

15 SIPRI Annual Report for 2002, see note 8.

16 You Ji, *The Armed Forces of China* (London, New York: I. B. Tauris Publishers, 1999), pp. 113-15.

8

China's Defence Industry, R&D and Defence Budget

Introduction

A country's overall economic performance, and central government's national budgets are closely connected to its defence expenditure and defence industry and, hence, these subjects are discussed together in this section. For a country that prides on the invention of gunpowder and rockets, the PRC's performance in the defence industrial field was dismal for most of the past fifty years. This was simply because the command economy and the Soviet model industry could not generate adequate surpluses or profits needed to build a vibrant defence industry. Mao suppressed domestic demand, fed his people Communist ideology and distributed poverty. He was successful in building a rudimentary nuclear deterrent which he used to great effect; but the defence industry languished even when it was able to make moderate exports.

It is only since the Deng visit to the southern coast in 1992 that the Chinese economy and industry have posted dramatic results. This spectacular achievement made it possible for the PRC leadership to allocate ever higher outlays for defence procurement, R&D and industry; a significant aspect of China's economic development for those studying its power projection capabilities. According to the first batch of foreign industry experts entering the PRC in the Deng reform period, evidence of abandoned steel smelting furnaces could be seen in almost every rural household; a direct result of Mao's dictates for transforming rural China into a self-sustained economy. Self-reliance was not just the national objective; it had become a dogma.

Through most of the 19th and 20th centuries, China had remained an inward looking self contained and self-sufficient and largely rural agrarian economy. There was little incentive for foreign trade and its experience was none too encouraging. In the first half of the nineteenth century, China encountered foreign ships bearing opium that it traded for tea and precious silver. Thereafter, foreign trade was largely the function of the demands of the five Western powers and later Japan that had extracted various concessions and created enclaves of influence inside China. The first half of the 20th century was lost to civil war and the Chinese peasantry subsisted with dogged determination and innovative multiple cropping to get more out of the unproductive land. China's population increased, as agriculture needed more hands but in the end, increased the burden on land and other natural resources. At the time of the liberation of the country in 1949, the state of its economy and industry was depressing. Mao had no option but to seek Soviet help to build China's economy and industry, which naturally developed on the Soviet model and philosophy.

As in other areas of China's national efforts, here too the developments can be clearly divided into three phases, the Mao era, the Deng era and the present era. Each era was characterised by a bold and unique experimental approach that was not always successful. During the Mao era, the Communist model imposed a rigidity that inhibited free and innovative thinking and yet Mao and his cohorts had wanted a distinct Chinese stamp. As expected, the emphasis was on heavy industry and technology that was already becoming obsolete. As discussed earlier, Mao then made a virtue of necessity and publicly denigrated the role of technology and highlighted the importance of the human effort. China could achieve anything; surmount any obstacle so long as the indomitable spirit of the Chinese people was constructively employed under the guidance of the Party's ideology. The twin goals of self-reliance and ideological supremacy in central planning put paid to any worthwhile improvements and both the economy and industry languished.

The arrival of Deng Xiaoping radically changed the national approach. Self-reliance was no doubt important but China's deliverance lay in foreign technology and financial assistance. Foreign Direct Investment soon became the buzzword and the PRC launched its open door policy and '*market economy with Chinese characteristics*' as its avowed goal. Defence industry and budgetary allocations began to increase as the national economy gradually gained strength and access to foreign technology improved.

Early Years

In early 1950, just months after its birth, the PRC's defence industry got a head start following the signing of the Friendship Treaty with the Soviet Union. By the end of the 1950s, it was manufacturing under licence, a variety of arms and equipment including rifles, artillery guns, mortars, tanks and even fighter aircraft. The Korean War experience had already highlighted the importance of military technology, the need to build the necessary industrial infrastructure and self reliance. Expectedly, China's defence and heavy industry was fashioned on Soviet lines and grew rapidly. The stress was on production of arms in quantity and not on quality or sophistication; as a result, by the end of the 1970s, the PRC had produced thousands of guns, tanks and aircraft which were already becoming obsolescent. These were meant both for its own armed forces and for exports to friends in the developing world. By supporting anti-Western revolutionary regimes in this way the PRC was also able to derive considerable political benefits even when it was internationally isolated.

By the time of the second coming of Deng in 1978, it had become clear that China's industry was in dire need of fundamental reform as it had decayed over the years due to uncertain policy direction during the turbulent periods of the Great Leap Forward and the Cultural Revolution. The only area that continued to receive attention and steady support was the part of the defence industry engaged in the development and manufacture of its strategic assets, viz. missiles and nuclear weapons. Even if less efficient, other defence industry continued to do what it knew best: produce obsolete and often sub-standard weapons. With the Chinese Communist Party leadership realising the critical importance of creating useable wealth for the enhancement of its overall national strength an attempt was made towards defence industry conversion which would allow China to utilise its vast Soviet style industrial infrastructure to produce civilian goods for internal consumption and export, thus helping economic growth. This experiment, however, was largely unsuccessful and resulted in the production of sub-standard and costly goods which were neither competitive nor was their adequate demand in the local and foreign markets.

Problems

Reform of State Owned Enterprises (SOEs) was the main agenda for the 15th Party Congress held in September 1997. The magnitude of the problem can be gauged by the fact that, there were some 305,000 SOEs, employing 70 per cent of China's urban work force (about 109 million workers) and generating 30 per cent of total industrial output.[1] Some of the major problems faced by China's Military Industrial Complex (CMIC) were state ownership, low productivity, and lack of competitiveness, commercial rigour and an excessive emphasis on self reliance when China was neither getting access to foreign technology nor did its rigid communist ideology permit joint ventures because of a deep seated xenophobia. There was also a general aversion to foreign ideas and innovation. The China Defence Science and Technology Commission (COSTIND) was expected to coordinate the "often contradictory requirements" of the PLA and the defence industry. It supervised the research, development, testing and evaluation (RDT&E) of all new weapon systems and equipment and set priorities, allocated resources and mediated the conflicting interests of the user and the developer/producer, but was not very successful.[2]

Reform of the industrial sector was directly dependent on the reforms in the financial and banking sectors. For decades, state banks had advanced loans to loss-making enterprises. But because the SOEs employed such a huge work force these could not be reformed or run on commercial lines without "painful and disruptive" adjustments.[3]

At the conceptual level, leftist thinking, adherence to self-reliance and aversion to foreign ideas and innovation plagued the defence industry. At the organisational level it was lack of flexibility and horizontal integration as it remained separate from the civilian and commercial sector.[4] In short, it was artificially propped up.

At the practical level too, the industry faced many difficulties such as scarcity of resources, lack of skilled experts, managers and labour. Problems of brain drain, poor infrastructure, technology absorption, dwindling markets and an absence of procurement doctrine added to its woes. Procurement procedure existed only in theory; the process that should normally include the development of annual, mid-term and long term planning guidelines, the tracking of developments in foreign weapons systems, analysis as to technological feasibility, operational requirements and funding levels, coordinated process of design, trial production, serial production and deployment but in practice, little thought was given to linking threat and strategy to equipment manufacture.[5] Even after a weapon systems had entered the engineering development stage, its operational mode and utility had not been determined. In sum, this ended up in the production of surplus and superfluous weapons and equipment.

For R&D to develop successfully, it needs to evolve along the phases of basic research, applied research, development, production, and finally product deployment. Unless organically coordinated, the whole process can suffer major setbacks. China faced many such problems.[6] China lacked such streamlined thinking, disciplined project development and coordination. In the last five years, the Chinese defence industry has made considerable progress and is now poised to make steady and possibly even remarkable progress in the next five to ten years. This optimism stems from the fact that China has been successful in accessing foreign technology such as aero-engines for the FC-1 and the J-10 fighter programmes and for joint ventures and licence production of modern systems such as the Su-30 multi-role fighter.

Defence Conversion

Started in the late 1970s, by 1992 more than 70 per cent of the production of civilian run defence industry under the State Council comprised civilian goods. Defence conversion was aimed at unloading decades-old ideological, organisational and technical restraints but this did not actually happen for the reasons cited above. The sixteen character slogan coined by Deng Xiaoping in early 1978, *"combine the military and civil, combine peace and war, give priority to military products, let the civil support the military,"* showed that the leadership did not appreciate the burden it was placing on the civilian industry by expecting it to support the ailing and inefficient defence industry. Defence conversion is usually aimed at utilising the experience and excess capacity of the defence industry to produce civilian goods as dual-use technology helps to keep military production lines warm, provides opportunity to develop military technology in a disguised form and allows rapid mobilisation for war. In China, defence conversion has been successful in areas where markets were immature like transport, high-end consumer goods like colour televisions and motor cycles but not in markets where there was competition. Weak management and poor market research leading to production of overpriced good already in the market place, inadequate economics of scale, lack of cost or customer consciousness, lack of financial accountability were seen as the biggest problems.[7] Defence conversion cannot be expected to provide a quick fix as it needs a radical change in the way the economy and industry are managed. For many years, the Chinese defence industry seemed unable to move beyond its planned economy mentality. China's defence industry success was often exaggerated while most factories in Sichuan were on the verge of bankruptcy. Defence conversion has not been particularly successful but whether or not it is totally stopped is difficult to ascertain.

PLA Withdrawal from Business

For about twenty years, from the early 1980s to the end of the 1990s, the Chinese Army operated the largest and most profitable military business complex in the world. The PLA invested enormous efforts and resources to build a vast but loosely-connected commercial empire stretching into virtually ever corner of the economy. The objective of entering business was to make up for the severe financial cuts imposed on defence spending in favour of economic development. In the absence of proper supervision there flourished corruption, smuggling, profiteering and other vices. These problems became so rampant that the political reliability of the military and the health of the Chinese economy began to be undermined.

It is estimated that at the beginning of the 1990s, at its peak, there may have been as many as 20,000 army-owned enterprises. Most of these entities were small outfits with only a core of around 500 to 1,000 large enterprises and conglomerates. An overwhelming proportion of the enterprises that were established during the late 1980s and early 1990s were small unprofitable ventures. Most military units lacked access to credit and there were only limited inflows of domestic and foreign investment. Even before its entry into business in the mid 1980s, the PLA was already economically self-sufficient. The logistics system had its own hospital, transportation and telecommunication system, factories and virtually any other operations that were required to support the running of a huge military machine.

Other PLA companies were lured into new fast-growing and lucrative but risky industries, such as real estate, speculative financial activities-futures and commodities trading. At a national anti-smuggling seminar held by the PLA high command in July 1998, Chinese President and Commander-in-Chief Jiang Zemin instructed the military, paramilitary and law-enforcement organs to clean up and divest themselves from commercial activities. He told the assembled military brass that "to properly develop the army in an all-round manner, the central authorities decided that the army and the armed police should earnestly screen and rectify various commercial companied operated by their subordinate units and shall not carry out commercial activities in the future."

The PLA reportedly, continues to retain substantial economic interests even after its separation from commercial activities. The divestiture may have reduced the size of the military business complex by upwards of 75 per cent in asset terms, but there are still several thousand small and medium-sized military-affiliated business entities in operation. Remaining enterprises are primarily concentrated in the agricultural, industrial and welfare-support sectors which are the same areas that the PLA was engaged in before the 1985 decision to go into business. The PLA farming system is being expanded to allow military units to partially compensate for the loss of commercial income through side-line production.

The PLA has kept control of several hundred industrial enterprises, maintenance and repair of factories and research facilities to provide logistical support for the rest of the military establishment. These factories produce uniforms, beddings, food and other quartermaster stores. Several thousand small-sized family-run workshops, shops and other entities have been allowed to retain their ties with the PLA to provide work and welfare support for family dependants and demobilised troops. These enterprises are generally located on or close to military bases. Military units also retained a significant proportion of hotels and guesthouses, although many were required to close their doors to commercial business and revert to hostels and guesthouses for state, party and military personnel.

The PLA has been permitted to retain a small number of corporations that are considered important for national security or for the public interest. This includes telecommunication and aviation, although it was forced to hand over a fledging mobile telephone network to the civilian authorities at the end of last ear. "The air force's China United Airlines has been allowed to continue to operate its key routes in China, which are usually to cities in impoverished outlying provinces."[8]

Future Prospects

In spite of many problems and setbacks, Chinese defence industry has posted impressive results in the last five to seven years. Due mainly to lack of transparency and poor access to information on China's defence and high-tech projects, it is common to underestimate China's capabilities. The Chinese industry is developing an all round capacity in high-tech areas of science and technology.[9] After absorbing the lessons of the recent conflicts since the 1991 Gulf War, the Chinese military and civilian leadership has realised that modern warfare is increasingly more dependent on the 'soft' side of technological and military capabilities such as information and computing, than the 'hardware' such as tanks and guns. Even when a country is not doing too well in the traditional areas of developments such as weapons design and manufacture, it could well pose a major challenge to a more capable adversary by developing its capabilities in the 'soft' areas. Very often, analysts have confused China's intent with the assessments of its capabilities and have generally underestimated the China threat.[10] China's culture of excessive secrecy makes it difficult to assess the true impact of recent changes in doctrine, strategy, arms imports, deployment and use of indigenously developed weapons, exercises, training, and logistics. Traditional indicators of military power or bean count cannot correctly reflect China's ability to wage and win modern wars. It could well harm or destroy enemy's information or communications infrastructure without having to resort to massive use of its firepower.

The so-called RMA technologies are so closely linked to civilian use and innovations such as the high powered personal computers, mobile phones, internet, hacking, virus attacks space-based telecommunications, command and control that civilian technology overlaps military application. For this reason, correct assessment of the size and the scale of the technological revolution that has taken place in China over the past fifteen years is becoming increasingly more difficult.[11]

From 1992, when China's economic reforms really took off, the PRC has been investing very heavily in computer, electronic, simulation, space and telecommunications industry. "Of the approximately US $40 billion that has been invested in China, US $7 billion has gone to the electronics and computer industry."

The spectacular growth of the home grown Chinese hardware and software industries is one indicator of China's major achievements in this field.. Between 1980 and 1990, China imported computers from the United States, Europe, and Japan, to the degree permitted by their foreign reserve balance and foreign export control regulations. As recently as 1990, there were only a few viable domestic computer manufacturers in China; the strongest of them, the Great Wall and Stone Corporations, were fighting to stay alive and improve quality, faced with the prospect of incipient market penetration by foreign firms such as IBM. In 1997, locally assembled computers accounted for 44 per cent of the market, and there are over 6,000 Chinese computer companies.

Wendy Frieman says, "While some of them are Mom and Pop shops, others, such as the Legend Group, are poised to replace IBM as the top seller of computer hardware in China. The emergence of the Legend Group is a particularly interesting illustration of the impact of market oriented reforms, reinforced by foreign investment and an infusion of foreign technology. This is a company that did not even exist until the post-Tiananmen economic boom. By 1995, it was in sixth place, behind foreign, as well as other Chinese, firms. In 1998, Legend sales increased by 80 per cent to 140,000 units, which put it in first place among Chinese competitors, thanks in large measure to two strategic decisions. One was an alliance with AST, a US manufacturer with a long time presence in China, and the second was the decision by Legend to open a design centre in Silicon Valley. Legend also negotiated an arrangement with Microsoft under which Microsoft loads a Chinese version of Windows on each of its machines."

Over five years ago, China had the world's fourth largest supply of Microsoft Certified Professional "SE trained technicians", the highest level of professional certification. Such is the enthusiasm for acquiring new skills that Chinese students routinely learn more in three weeks than what Microsoft training executives thought would take six. Microsoft was planning on training over 200,000 people in China at its Authorized Technical Education Centres, and then on providing incentives for top customers to recruit exclusively from those centres. IBM had also announced a donation of $25 million to establish customer training centres in conjunction with China's top technical universities, over and above its commitment to invest $2 billion in China over the next five years. The Chinese version of Office 97 was available only three months after its release in English in January 1997.[12] In the late 1980s, the Chinese chip industry, using imported technology, was over ten years behind the Western state of the art. Wafers were three inches in diameter and line widths were somewhere between 5 and 10 microns. It was thought that China would take a long time to catch up. But by 1995, the industry in China was producing 125 mm line widths (against a world standard of 200mm) with 1-3 micron features (compared with US standards of .35 to .75 micron features). And in May 1997, China announced that NEC Corporation would invest $1.2.billion in a chip factory that would produce 8-inch (200mm) wafers with .30 micron line widths. This has put China very close to world standard, especially surpassing the pace of progress in the US and Japan.

Telecommunication technology was identified early in the modernisation programme as a critical bottleneck for China, and one that would take several decades to update to Western standards, and this has paid rich dividends. One relatively new entrant is planning to spend $11.5 billion building a network to support 12 million digital exchanges. What with widespread use of fibre-optics line networks, China will comfortably leapfrog telecommunications technology and catch up even faster with the Western world. This has enabled rapid growth in Wide Area Networks in China, many of which support higher education and advanced scientific research, as well as commercial programs. Most are accessible via the Internet. Since 1994, when fewer than 10 per cent of Chinese PCs were part of a wide or local area network, the growth in demand for routers, hubs, and switches, has grown by 150-200 per cent a year.

Satellite based navigation, communication, and surveillance constitute another arena where Chinese capabilities developed for civilian purposes could become critical success factors in future conflicts. China's commitment to the development of satellite technology and space sciences is currently focussed on commercial and civilian applications, especially telecommunications and environmental monitoring.

Since 1970, China's Long March rockets have launched fifty satellites with an 88 per cent success rate. Despite some failures in 1996, the programme appears to be back on track. In March 1998, a Long March rocket successfully launched two US-owned Iridium satellites; China announced at this time that its space company, Great Wall, had, captured 7-9 per cent of the international market between 1990 and 1997. China's rockets have been used to launch satellites in geosynchronous orbit, as well as low-to medium-earth orbit, and there were plans for at least ten launches before the year 2000. US and other suppliers are worried that assistance in the civilian field would ultimately benefit not only China's space launch capability, but also contribute to enhanced missile accuracy. As per a leading aviation magazine, "China launched a Long March 4B from Taiyuan on 15 May 2002 carrying the 428kg Fengyun1D meteorological satellite and the country's first dedicated ocean monitoring satellite the 360kg Haiyang 1 into polar, sun-synchronous orbit. China plans to launch the biggest ever thermo-optical/X-ray space telescope in 2005"[13]. It was also reported that China had postponed its plans to launch a national direct broadcast satellite indicating its national priorities.

The rapid growth of the space launch industry occurred at roughly the same time as the dramatic increase in China's ballistic missile production. Although most of the improvements came in the area of medium and short range missiles, satellite guidance and telemetry are obviously more helpful for long range ICBMs. Even so, the connection between civilian and military uses of technology is noteworthy.

"One example of China's ambitions in this area is the recently announced joint venture with Brazil to launch a satellite for environmental monitoring. The China Brazil Earth Resources Satellite project was scheduled to buy a satellite ground station from Hughes STX in Maryland. China's acquisition of two satellites designed to use synthetic aperture radar from Spar Aerospace of Canada also points in the direction of high-tech growth areas. The SAR satellite provides China with much more accurate and detailed images, as microwave instruments can acquire and transmit signals through cloud cover, darkness, and severe weather conditions. Although the purpose of the acquisition was announced by the Chinese as monitoring natural resources and disasters, one of the satellites will be built specifically for COSTIND. As we shall see later, China's 1998 forecast launch of a manned spacecraft in the near future have already fructified".

From about the beginning of the national economic and technological reforms the Chinese leaders recognised that imported technology was no doubt essential but development of human intellectual resources would be more difficult and time consuming. Technology imports can only provide a short-term fix; "an indigenous base of scientific and technical talent is essential to China's rise as a great power on the world scene."

"The number of Chinese students and scholars in the United States reached a high watermark of 55,000 in 1992-1993. Of these, 53 per cent were studying physical and life sciences, engineering, or math and computer science; an additional 17.5 percent were studying health sciences, business, or agriculture. Statistics about the numbers who are returning to China are not available, but an informal consensus exists in the international education community that it is approximately one-third." Another American analyst puts the number of Chinese students in the West at 140,000.[14]

Worries of brain drain have also proved exaggerated with one in three students returning to China. China has certainly compressed, even leap-frogged the time taken to absorb sophisticated technology and it is evident that the credit goes to "the return of smart and dedicated people rather than to the purchase of expensive scientific instrumentation."

Even those continuing to live in the US and Europe maintain regular contact with their friends, colleagues and universities enabling to *serve as middlemen and interpreters between Chinese and Western science, and to transmit knowledge and ideas back to China.* Some of this might be occurring through intelligence channels, to be sure, but much of it is happening through market-driven economic activities. Those who stay feel a strong obligation to help their colleagues back in China in whatever way they can and, thanks to the rapid expansion of the telephone system in China and the growth of the Internet, they are able to maintain regular communication with communities back in China. "It is not unusual for PRC scientists in the United States to exchange email with Chinese contact several times a day."

Some of the students are also raising funds to build institutes for conducting high-tech research back home and have helped their 'alma mater' to develop and maintain links abroad. As a result of the overall internationalising of the PRC, the Chinese system has become more flexible and open further encouraging access to and quick absorption of hitherto inaccessible dual-use technology. Only 4 per cent of China's college-aged population attends a university. China graduates no more than 110,000 engineers a year, and fewer than 10,000 math and computer science majors. Expenditure on education is only 2.2 per cent of GDP, compared with 22.3 percent for Singapore or 15.9 percent for the Philippines. But despite the difficulty of getting students to come home, China has made considerable progress.[15]

Despite the PRC government's avowed interest in promoting science and technology since 1978 Chinese science remained in a fairly dismal state for the next ten years. The PRC government, however, succeeded in depoliticising science during this period, "and by the late 1980 it was rare to find that an individual's position on a substantive scientific issue would bring with it ideological baggage. Salaries and living conditions improved incrementally. Some foreign exchange reserves were allocated to universities and research institutes for importing computers and other equipment. Tiny pockets of excellence began to emerge, specially in fields targeted by the central government for rapid development. But it was not uncommon during this period for foreigners to visit deserted laboratories where no one seemed to be working and where imported equipment was both underutilized and poorly maintained."[16]

The aim of the Programme '863', codenamed after the date it was launched, viz. March 1986, was to identify and fund high-end research projects at the leading edge of scientific or technical knowledge to keep pace with international high technology and to strive for breakthroughs.[17] "The concentration of resource is not a new concept for China: in earlier periods the government had identified 'key' laboratories and 'key' universities for targeted subsidies. But recently, the number of investment targets seems to have been reduced even further, with the Chinese Academy of Science having selected ten major science projects, the '863' program having identified five fields for concentrated investment, and the State Science and Technology Commission having specified fifteen key science projects. These projects or key areas include many of the 'usual suspects': high performance computers, a synchrotron radiation accelerator, gene injection technology, computerised manufacturing, and so forth. And it is worth noting that, for the most part, *they are not identified specifically enough for outside observers to be able to systematically track their progress over time.*"

Here is another success story. "The 'True God Advanced Computer Company', a profit-making subsidiary of the Shanghai Academy of Sciences Computer Centre, began in 1986 as an internal service organisation to support other branches of the Academy and quickly expanded into business areas that could earn profits to support scientific research in an era of declining subsidies. Among the projects completed by True God are, a colour-matching system using photo-spectrometer and PC software, an animation software package for television commercials, an automated toll collection system, and a debit card system for use by taxi passengers."

Other examples of collaboration between academia and their subsidiaries include the Keli Corporation, an offshoot of the Chinese Academy of Sciences Institute of Acoustics that develops products for the oil and gas industry, San Huan, tied to the CAS Institute of Physics and engaged in production of magnetic materials, Kejian working on phased array scanning systems for clinical diagnosis in medicine, Daheng attached to six different optics institutes of the Chinese Academy of Sciences and engaged in lasers and optical fibre production, and Hefei's Science Island, an industrial park near Hefei's University of Science and Technology, with numerous high-tech firms that are commercialising university-developed technology.

The major reason for these sweeping and radical changes in the approach to the development of defence industry was once again China's fortuitous foreign contacts. William Perry, the then US under-secretary for defence gave some sane advice to the Chinese, especially Gen. Zhang Aiping in 1980. This was the period of normalisation of relations between the US and China. "Perry concluded a 1980 visit by bluntly informing the Chinese that their ambitions were unrealistic. Broad-based defence technology co-operation was not feasible in the areas that Beijing had chosen simply because China's low technical levels, and the utter backwardness of its R&D system would pose serious inter-operability problems". Instead, Perry suggested a longer-range, less military-specific, more broad based program of technology cooperation that would fundamentally improve China's S&T field. China's real need was to build a national technology infrastructure which, if handled well would give the Chinese some dual-use potential. A purely military technology base, Perry added, would simply be too narrow a foundation upon which to modernize China's industry, much less to overhaul its backward conventional armed forces. The PRC was thus able to first make a sound scientific and industrial base before undertaking research and development in truly high-tech areas. Instead of borrowing complete systems from the US, China shifted its focus to components or sub-systems that were of wider industrial use and to software-based processes that were needed across the entire modern industry. In short, contacts with the US experts in the 1980s helped guide China's policy makers to determine the broad direction and methodology needed to take their defence industry to a completely new and hitherto unknown levels of sophistication.[18]

Another successful field of great value to the PRC's power projection capability has been the military aviation sector. Here too, the PRC experienced innumerable roadblocks for decades. These included a general lack of understanding of air power employment and, hence, the type of weapon platforms needed for modern air warfare, little experience in project planning, non-availability of engine and avionics technology and finally scarce resources. Thanks to China's painstakingly cultivated contacts with the industrial West and its fast burgeoning economy, many industry majors in Europe and the US continued to give invaluable support even after the 1989 Tiananmen ban on transfer of technology. It is easy to imagine the power that China wielded in global markets once annual FDI into China had jumped from US$ 10 billion in 1991 to US$ 40 billion by 1996. In the post-Cold War atmosphere, not only were countries like Russia ready to sell China the latest technology but were keen to maintain their sales–even the Europeans were eager to jump on multi-country joint ventures especially avionics support that faced few export problems as they fell in the so-called 'non-lethal' category. The steady growth and incremental modifications and upgrading of the J-7 (MiG-21) variant from a vintage interceptor into a relatively modern, manoeuvrable and yet affordable fighter capable of delivering air-to-air missiles and up to 3,000 kg of armament is a case in point. In the recent past, China sold some 50 F-7PGs with an Italian Grifo airborne radar to Pakistan and is now modifying over 100 fighters in the Pakistani inventory. Further, with the assured supply of two very successful Russian engines, the Klimov RD-93 and Saturn Lyulka AL-31F, the PRC is well on the way to produce large numbers of FC-1 and J-10 fighters that have been languishing at design level for many years. Such is the pull of money and success that many European and American commercial and military aviation majors are only too ready to give China what it demands.

China achieved a major breakthrough on 15-16 October 2003 by successfully launching a manned space mission. Although, some sceptics said it was not a unique or even major technological breakthrough or a scientific innovation, the PRC had reached a significant milestone in its ambitious space programme. The PRC has once again proved, if proof was indeed necessary, that it has the determination to make good on its promises, whatever the difficulties. China's defence industry is likely to continue its steady growth in short range ballistic missiles, cruise missiles fighters or unmanned air vehicles and air-to-air missiles and associated avionics so that in another five to six years it could well become a formidable competitor in low-cost high-tech weapon systems and aircraft.

PRC's Defence Budget

A broad overview of the PRC's economic indicators would be useful to get a clear perspective of its defence expenditure. From almost a subsistence economy until the end of the Mao era, the 1978 Deng reforms began to slowly change the very outlook of the Chinese decision-makers on ways to create wealth. Deng said, "By self-reliance we mean that a country should mainly rely on the strength and wisdom of its own people, control its own economic lifelines, and make full use of its resources." He found nothing wrong in selectively importing advanced foreign technologies and paying for them by increasing exports of arts and crafts products, industrial and mining products and resorting to standard international economic practices such as long-term trade credits and other forms of foreign assistance. In short, China was no longer averse to opening its economy to global market forces and ideology would no longer be allowed to impede economic progress. Although some of his colleagues called him a "capitalist roader", currying favours with the capitalist world, he argued that increased foreign exchange earnings would pay for the import of modern equipment and technology so essential for industrialisation and overall economic progress.[19] As a result, China's foreign trade grew at an annual rate of 15 per cent from 1984 and reached US$ 281 billion in 1995. The PRC had already quadrupled its 1978 GDP by 1990. Annual Foreign Direct Investment also grew at an impressive rate from about US$ 10 billion to US$ 40 billion by 1996 and crossed the US$ 50 billion mark by 2003. China's foreign currency reserves stood at an impressive US$ 400 billion in 2003 and its GDP at close to US$ 01070 billion. By the mid-1990s, the PRC had at last begun to generate the extra funds needed for a comprehensive modernisation of its armed forces.

As a rule, defence budgets are meant to convey to the outside world the fundamental philosophy of the state towards national defence and assure it that the country is only trying to meet its legitimate defence needs. But the PRC's proclivity for excessive secrecy with the country's defence and strategic activities does not allow an outsider to come to any accurate estimate of its defence spending. At the same time, differing 'guesstimates' from different quarters make the problem even more complex because until 1998, when the PLA was officially banned from running business enterprises, profits generated from these business ventures were routinely used for funding defence requirements. The PRC leadership is unlikely to become more transparent about its defence outlays in the near future and we will continue to see some pro forma details of defence expenditure in the official statements. Consequently, its defence outlays have also been a subject of intense speculation for many years. Western analysts have consistently questioned the truthfulness of the official defence budgets of the PRC and their estimates range from two to four times the official figure of around US$ 20 billion. It is, however, likely that in a few years, China's overall economic growth and the consequent clout and power that it will wield, would become so formidable that such debates would have little value, and become merely academic. The PRC has acknowledged in its official statements, including the 2002 Defence White Paper, that there have been small increases in its defence outlays but has attributed them to the growing needs of modern defence and have termed them "compensatory" in nature.[20] By comparing its defence outlay with those of many advanced countries, the White Paper attempts to prove that the PRC's budget continues to be the lowest in terms of the share of the GDP and that of central government expenditure.

To be sure, the Chinese defence budget had indeed declined in the first decade of the "four modernisation's" programme when resources were directed to other areas of the economy. The consistent double digit increase in its budget over the last decade, however, has caused concern the world over, but the PRC continues to justify these increases as legitimate needs of defence modernisation. The Official Defence Expenditure also does not give any details of the allocations to various sub-heads except in very general terms.

According to Arthur S. Ding, the Chinese also tend to use two different terms, viz. defence and military expenditure, further compounding the confusion. "In 1995 the Ministry of Finance divided the budget into two parts: central and local. The total amount allocated to defence expenditure was RMB 63.097 billion (US$7.38 billion), of which RMB 62.807 billion came from the central budget and the remaining RMB 290 million from the local budget."[21] It was not clear why this small portion of the budget came from local sources nor was there any indication about how this amount would be used. It is somewhat unusual for centrally controlled and managed organs of the government such as the PLA to be funded by local sources. This practice has apparently been discontinued as the 2002 White Paper on Defence does not split the budget amount.

The White Paper gives five reasons for the 'small' increase in the defence expenditure. These are: (1) Increase of personnel expenses. (2) Establishment and gradual improvement of a social security system for servicemen. 3) Increase of expenses of a maintenance nature. 4) Increase of expenses spent on cooperation with the international community in anti-terrorism activities. (5) Appropriate increase of expenses for the improvement of military equipment to enhance defence capabilities under the conditions of modern technologies, particularly high technologies.[22]

While increases in the budget to cater for higher pay and allowances to the PLA personnel are natural and even necessary, additional expenditure due to "China's participation in international anti-terrorism activities" is not convincing. The increase of 84 per cent salary raise for officers and 92 per cent rise in allowances of soldiers are also an indication that the PRC can no longer ignore the legitimate demands of the PLA especially after its exit from lucrative business enterprises. The White paper goes on to add, "Since the armed forces stopped commercial activities in 1998, the budget for training has increased year by year." Increase in maintenance and automation have also been cited as another reason. This is an official admission, if an oblique one, that the PLA was able to top up its funding needs through its business profits.

The 1995 White Paper on National Defence had divided China's defence budget in three categories, viz. personnel expenses, maintenance and equipment, and the allocations were 36, 33 and 31 per cent respectively.[23] There is again no mention of R&D and weapons procurement, or capital expenses which are, "shown under various other ministries and departments or projects. The PLA also benefits from all kinds of subsidies and other incentives and has been running a huge business enterprise; from pig farms to five-star hotels." Jasjit Singh points out another interesting anomaly: the PLA's budget increased when there was a huge reduction in its overall strength but increased when the strength actually decreased with Deng's reforms cutting it down by one million soldiers. It is this inverse relationship that gives rise to doubts about the official figures Despite these anomalies, however, there is no gainsaying the fact that China's defence expenditure had been rising at an average annual rate of 15.83 per cent during 1992-98. This according to the same source is 32 per cent higher than the inflation during the same period.[24]

According to the 2002 White Paper, "In March 2001, the PLA began to reform the budgetary planning system. Drawing on the internationally adopted budgeting method, the PLA reformed its budgeting form, method and content extensively, with emphasis placed on implementation of the zero-base budget method." There is a noticeable change in the language, indicating a possible shift towards a more acceptable and transparent form of budgetary model.[25] This also shows that the laws, regulations and statutes instituted in the last two years are helping to regularise the Chinese system. The PRC leadership is perhaps more confident about managing its finances and, hence, does not mind sharing a few more details with the world.

The PLA modernisation drive was, however, not funded by the official budgets alone. Till recently, it used to get a generous share of the profits that the PLA made by running civil businesses and from the exports of arms. "The profit from these ventures, at the last known estimate, was $5 billion per year". China is both an importer and an exporter of arms."[26] While imports comprise high-tech weapon systems and equipment, the exports have largely been of low tech items but included fighter aircraft such as the J-7 and F-7P (MiG-21 variants), Jianghu frigates and tanks, mainly to developing countries such as Pakistan, Bangladesh, Iran, Libya, Iraq and many others. Sizeable funds were, thus, available from arms exports.

It is evident clear that the PRC's actual defence expenditure is considerably more than the officially declared figures. As we have seen in earlier sections, China had already quadrupled its 1978 GDP by the end of the last century; and its economy has continued to grow at an impressive rate. The value of China's trade had jumped from a paltry $45 billion in 1980 to $470 billion by the year 2000;[27] showing the great surpluses that its economy must be generating to support military modernisation. The PRC would, thus, find more than adequate financial resources to maintain the pace of its military modernisation.

As noted earlier, defence spending in China has doubled in real terms between 1990 and 2000. In 2002, the country's official defence spending was US$ 20.07 billion. Total defence spending is reckoned to be much higher, with estimates ranging from US$ 35 billion to US$ 65 billion.[28]

Assessment

We have seen that the PRC's defence industry has reached a take off stage in many sectors. With the added emphasis on high-tech sectors, it is quite likely that China would move even more quickly towards manufacture in the RMA technology area. The growth rate for its GDP has been estimated at a staggering 9.0 per cent for 2003. Impediments like SARS, the Iraq War and a general down trend in global economy have not adversely affected China's growth rates. It is likely that the PRC leadership would continue to make generous allocations for defence in the next five to ten years and this will have a salutary effect on the PRC's capability to project power in the region.

Notes

1 Harlan W Jencks, "Costind is Dead, Long Live Costind! Restructuring China's Defence, Scientific, Technical, and Industrial Sector", in James C. Mulvenon and Richard H. Yang (Eds.), *The People's Liberation Army in the Information Age* (Santa Monica: RAND Corporation, 1999), p. 60.

2 Ibid., p. 61.

3 India too has faced similar problems with its public sector enterprises; no one wants to buy the loss-making and inefficient ones while the government is wary of selling the better performers.

4 John Frankenstein and Bates Gill, "Current and Future Challenges Facing Chinese Defence Industries", in David Shambaugh &

Richard H. Yang, Eds., *China's Military in Transition* (Oxford: Clarendon Press, 1997), pp. 131- 163.

5 Ibid., pp. 143-151.

6 Ibid., p. 144.

7 Ibid., pp. 153-154.

8 Timothy Hu Reports "Forces in Business" *Janes Defence Weekly* (JDW), 14 February 2001, p.23, also see Jencks, n. 1, p. 73.

9 Wendy Frieman, "The Understated Revolution in Chinese Science and Technology", in James R. Lilly & David Shambaugh (Eds.), *China Military Faces the Future* (Washington DC: American Enterprise Institutefor Policy Research (AEI) 1999), pp. 247-267.

10 Ibid., p. 247.

11 Ibid., p. 248. All the quotes are from Wendy Frieman and the information is also taken from the same source.

12 Ibid., p. 252.

13 *Flight International,* 28 May-3 June 2002, p. 30.

14 Andrew J Nathan & Robert S Ross, *The Great Wall and The Empty Fortress-China's Search for Security* (New York London: W. W. Norton & Company, 1997), p. 171

15 Also see David Zweig, *Internationalising China-Domestic Interests and Global Linkages* (Ithaca & London: Cornell University Press, 2002), who also focusses on China's academic exchanges with the West to asses its internationalisation.

16 Frieman, n. 9, pp. 157-159.

17 For more details, see Akshay Joshi, *Information Age and India* (New Delhi: Knowledge World and IDSA, 2001), pp. 75-140.

18 Evan A Feigenbaum, "Who's Behind China's High-Technology Revolution?", *International Security,* Vol. 24, No. 1, (Summer 1999), pp. 104-106.

19 Nathan & Ross, n.14, pp. 160-61.

20 China's Defence White Paper-December 2002, available at FBIS/CH dated 12-10-2002, p. 14.

21 Arthur S Ding in David S Shambaugh & Richard H. Yang, Eds., n.4, p. 20, 165.

22 White Paper, n.20. p. 14.

23 Jasjit Singh, in *Asian Strategic Review* (New Delhi: Institute for Defence Studies and Analyses, November, 1999), p. 40.

24 Ibid., p. 40.

25 White Paper, n. 20, p. 14.

26 Singh, n.23, p. 44.

27 Zweig, n.15, p. 10. Though somewhat technical, this is an excellent study of how and why China internationalized its economy.

28 "China's PLA Modernisation and Regional Concerns", *Asian Defence Journal*, (July and August 2002), p. 6.

9

Implications for India's Security

Today, China's economic power makes US lectures about human rights imprudent. Within a decade, it will make them irrelevant. Within two decades, it will make them laughable. By then the Chinese may threaten to withhold MFN status from the US unless we do more to improve living conditions in Detroit, Harlem and South Central LA

Richard Nixon 1994[1]

Introduction

It is now ten years since Nixon prophesied about China's growth. His forecasts were remarkably close. He had predicted China's economic growth at a time when its inflation rate had reached an all-time high of 23 per cent.[2] That the PRC was successful in overcoming the roadblocks to its open-door reforms programme speaks volumes for the determination and perseverance of China's Communist leadership. It becomes even more praiseworthy when seen in the context of the 1989 student demonstrations at Tiananmen, and the speed with which they were brought under control, albeit with excessive force. The history of the last fourteen years since the disturbances, shows that the central leadership was largely successful in containing the fallout of the students' protests without seriously harming the pace of economic reforms. As we shall see later, the 'fourth generation' (Disidai) leadership is, if anything, even more adept and focussed on taking the reforms process forward.[3] China's growth curve may have flattened somewhat but its learning curve has undoubtedly become equally smooth. It is evident therefore, that the PRC's power projection capability would be formidable by 2010.

In Chapter 1, we saw how the PRC had skilfully managed its foreign relations so that by the time the Cold War ended, China was once again ready to do business with Russia the successor state of the former Soviet Union with whom China had gone to war. We also examined in Chapter 2, the propensity of the PRC leadership to use force albeit with an eye on escalation control and utilisation of propitious opportunities when these presented themselves. We saw in Chapter 3, the evolution of the PLA's doctrine and how it slowly but definitely emerged from under the shadows of Mao's 'people's war doctrine'. In Chapter 4, 5, 6, 7, we focussed on how the PRC's quest for comprehensive national power had, in fact, radically transformed its nuclear and conventional forces. Chapter 8 discussed in some detail, the PRC's 'thrust areas' in the choice of high technology and the consequent emphasis on RMA. This study would now turn to the implications of China's fast burgeoning power, for India's security.

The PRC's quest for power is steadily bearing fruit. This has been possible due to its dramatic achievements in the economic field. China today has fast growing economic reserves that permit generous funding of its military modernisation in myriad ways. Further, China's overall political and strategic strength along with its membership of the UN Security Council, and its position and prestige in the comity of nations, confer on it the unique ability to ward off threats to its security while at the same time ensuring a relatively benign international environment necessary for its economic growth. In sum, China has not only neutralised all external threats but its power evokes awe and respect even from the only super power. By selectively resorting to force, the PRC has also reiterated on a number of occasions its resolve to discourage any country from taking undue liberties with it. In this chapter, we shall examine the likely ways in which China might use its power to constrain its competitors and rivals.

The Strategic Setting

In the past, the PRC has revised its foreign policy and national security perspectives on at least three occasions. Its alignment with the former USSR was to last just over a decade when relations soured over ideological leadership and nuclear assistance issues. In the 1960s through 1970s, the change in its foreign policy was necessitated by its fear of the Soviet threat and the need to balance it by coming closer to the US. In the early 1980s, Deng Xiaoping felt that although the international system remained chaotic, the threat of a major war had evaporated. He also did not want China to operate from a position of a junior partner of the US. He, therefore, laid greater stress on economic development and began to improve relations with major and minor powers in an attempt to create a peaceful environment that was conducive to China's overall development and growth. It is now clear that the real aim of the 'four modernisations', in which defence was given the lowest priority, was, in fact, to build its national power so that China could achieve its long cherished goal of becoming a global power. Unlike in the first twenty years after the foundation of the PRC when its military was decidedly defensively oriented, the recent attempts at military modernisation are markedly offensive. Mao had no doubt emphasised violent class struggle and revolution but had relied more on ideology and diplomacy than on armed strength. In any case, the PRC simply did not possess the necessary financial and technological resources to build a modern military apparatus. As a result, the PRC unnecessarily built a huge military, based on increasingly obsolescent equipment that was never quite fit to wage a modern war. As we have already seen, the recent attempts at military modernisation have been more deliberate and focussed on enhancing China's ability to influence events abroad.

An important and perhaps the most demanding preoccupation of the current PRC leadership is the management of its relations with the US. Just as American scholars advocate a policy that combines 'containment' and 'engagement' to manage a rising China, the PRC too consciously promotes its relations with the US without giving in to the latter's dictates. US military presence in Asia Pacific, its special security relationship with Japan whose military spending is watched with concern in China, and finally the American backing for a status quo in Taiwan are perhaps the most compelling challenges facing the PRC.

China has thus far managed, through a combination of economic inducement, and military threats, the prevention of Taiwan's drift toward independence. "One potential crisis flared briefly on August 3, 2002, when the president of Taiwan, Chen Shui-bian, said he would urge the Taiwanese legislature to consider passing a law to govern the holding of a referendum on the island's future status. Since Beijing considers holding a referendum tantamount to declaring independence (regardless of the referendum's outcome), the statement raised the possibility that China might go to war over Taiwan in the near future......But Chen quickly moderated his rhetoric and Beijing was given Washington's assurance that.it did not support Chen on the issue. By late August the issue had died down."[4] It will, thus, be seen that management of the Taiwan issue is intimately linked to US actions in the region. It is also safe to assume that the US too would continue to periodically test the Chinese resolve by creating situations that force China on the back foot.

Japan's possible rise as an independent military power in step with its economic might is also an important factor in PRC's foreign policy calculations. Many Chinese believe that despite Japan's pacifist constitution, the voices favouring a more robust and independent strategic role for the island nation may pose a threat to China. Paradoxically though, continued American military presence in the region as a senior and restraining partner in the special security relationship with Japan suits the PRC.

The PRC has also voiced its unhappiness over what it sees as US hegemony and has vowed to work for a truly multi-polar world order based on sovereign equality.[5] China, however, selectively supports US moves and policies on nuclear non-proliferation, terrorism, narcotics trafficking, without conceding on trade, environmental and human rights issues. Having developed a mutually beneficial strategic relationship with Russia, China is often able join forces that oppose US unilateralism. So long as, Russia meets its needs in military equipment and technology, this relationship is likely to remain sound. China's attempts to make new inroads into the oil and gas rich Central Asian region have not been entirely successful. The long-term American military presence in the troubled areas adjoining Afghanistan following 9/11 would remain a cause of concern for the PRC. The recent Indian forays into Tajikistan for renovating a military airbase near Dushanbe and the implications of Indian military presence in the region are bound to cause some consternation in China.[6]

During the 1960s and 1970s, the PRC supported anti-capitalist revolutionary regimes in their fight for world revolution, but found few useful friends. From the 1980s onwards, the PRC toned down its revolutionary rhetoric but continued to export arms where possible and necessary to restrain the US from giving sophisticated arms to Taiwan. Simply stated the policy was that, "if the US arms Taiwan then the PRC proliferates nuclear weapons and missile technology to third world countries such as Pakistan, Iran, Saudi Arabia and Libya". Today it attempts to foster its ties with European and other countries on a much wider basis that includes, trade, technology transfer, political cooperation in the UN and its various sister organisations and through one to one diplomatic contacts. The major aim of China's foreign policy is to build a large constituency as a possible counterweight to the US. China's trade and scientific technological cooperation with the UK, Italy and many other countries in Europe is quite extensive and is aimed at creating a solid opposition to US unilateralism.

For some time now, the PRC has been managing its relations with the countries of South-east Asia by playing a more active and constructive role in the various multilateral fora such as the ASEAN, ARF and APEC. By signing a code of conduct declaration in November 2002, for the peaceful resolution of South China Sea disputes, China may have assuaged the fears of the smaller countries of the region, but the memories of China's 1974, 1988 and 1995 seizures of various island territories (Xasha, Nansha and Mischief Reef) cannot be forgotten so easily. In 1996 again, China had clashed with the Philippines over the Scarborough Shoals claimed by the latter.[7] China's fast growing economy has already affected the region's overall balance of economic power with the PRC calling the shots. Most ASEAN members thus continue to see China as a long-term challenge, but also see major opportunities in China's economic growth. They try to use the multilateral fora to manage the growing influence of China. As a leading strategic analyst in the region Amitav Acharya has said, "Notwithstanding many contentious territorial and economic issues in their relationship, the Southeast Asian countries have attempted to manage the growing power of China by eschewing a confrontational policy."[8] With its fast growing maritime power, China is quite likely to play a more active role in controlling piracy, and protecting the Sea Lanes of Communication (SLOC), but in the short run, its scope would most likely be limited by regional sensitivities.

South Asia, on the other hand, is an area that affords much scope for Chinese assertiveness and even adventurism. American and Japanese threats can be managed to some degree. The threat of Russian expansionism has long diminished if not altogether vanished. India, though enjoying a lower priority in terms of immediacy, continues to grow in a way that it could well become a challenger to China in the future. Managing and curtailing Indian power is, thus, a necessity from the Chinese viewpoint. This China has been doing by developing multifaceted bilateral relations with India's smaller neighbours. Interestingly, all these small countries share their borders with India but not with one another and hence have few disputes among themselves. The only problems or disputes that they have are with India, a fact that Beijing tries to capitalise on.[9] Pakistan, Myanmar (Burma), Bangladesh and to some extent, Nepal are easy targets for political manipulation. All of them desire a relationship with the PRC as a counter weight to perceived Indian hegemony. This allows much scope for mischief and it is clear that the PRC would take full advantage of the situation in constraining India's options and strategic space. Therefore, China's burgeoning military and economic power poses a more serious challenge to India than to any other country in the region. Another reason why India is likely to remain in China's sights is its role as a possible counter weight to China and the Russian and American attempts to use India in such a role. India is the only unattached large country with major power potential that the West may woo to keep China under check. Whether or not India accepts such a role is another matter. While the Indian decision making elite is no doubt aware of this long-term Chinese challenge, it is unable to do very much to protect India's interests due mainly to economic hardships and a fractious polity.

Strategies

At a minimum, China wants to eliminate or at least weaken any major rival while consolidating its power but that does not necessarily mean that it would readily and/or indiscriminately resort to use of military force. As Valerie Niquet of IRIS, Paris has said, "...the main objective of Chinese leadership is to establish the right to control and influence; and to be able to see this right recognised by all major powers of the world."[10] Wu Xinbo echoes similar sentiments in a slightly different format, "The PRC understands that the best way to defend its interests in the international political, economic and security realms is to make its own voice heard in the rule-making process."[11] Writing about the PRC's diplomatic manoeuvrings in creating a favourable international environment for its domestic economic development, Wu Xinbo says that the PRC follows different strategies. These include (a) freedom of action so that it can adopt an independent diplomatic posture; (b) maintaining a favourable balance of power in the world arena by developing close political or strategic relations with real and potential adversaries of the US; (c) isolating the US by exploiting contradictions among its allies – for example, using economic incentives to lure Taiwan, Japan and Western European countries to pursue China policies different from that of the US; (d) balancing US influence by forging strong ties with the Third World states by defending their rights and interests and through economic aid, and by encouraging multipolarity. China also favours bilateral settings for its dealings with other nations as it finds such interactions free of outside influences. Finally, by following what Wu calls 'omnidirectional diplomacy' China is trying to make as many friends as possible but adopts a differentiated approach depending on the power, size and the stage of development of the interlocutor concerned.[12] China, however, has not been entirely successful in winning new friends, but as a rising power, what concerns the PRC is compliance out of fear if not out of respect. India needs to understand this trait.

Its oft-repeated national objectives of reunification and of simultaneously enhancing its Comprehensive National Power (CNP) implicitly include the threat to use force and, hence, need to be taken seriously. In reality, reunification means the physical control and assimilation of all those territories that China lays claims to, irrespective of their present legal status. To do that, China must want to weaken the US alliance system in the Asia Pacific region. Acquiring the tools of power projection is integral to that strategy and so is creating new economic leverages.

The most recent challenge to the PRC's foreign policy establishment came from an unexpected quarter. On 7 November 2003, the Government of Kiribati, a tiny island in the Pacific, announced the establishment of diplomatic ties with Taiwan despite having friendly relations with the PRC. China has disallowed any country to have formal diplomatic relations with Taiwan, which the PRC considers a renegade province. Unless a country expressly endorses the PRC's one China policy, it cannot retain diplomatic relations with the PRC. Besides lodging a strong protest, the PRC also warned the Kiribati government, "[To] correct its wrong decision immediately, otherwise it must take full responsibility for all serious consequences." China also blamed Taiwan for what it called its 'unscrupulous policy of money diplomacy' to create two Chinas or one China and one Taiwan.[13]

The problem apparently began with the Kiribati Opposition party alleging the PRC's interference in local elections with the Chinese ambassador paying a donation of a few thousand US dollars to one presidential candidate. The PRC has enjoyed good relations with this island nation state and has established a space station on the island; that was used during the recent manned flight into space. It would be interesting to see its reaction as the PRC has not brooked any challenge to its one-China policy. The PRC obviously has three choices, first, downplay the problem and somehow get Kiribati to reverse its decision through inducements, second, break off diplomatic relations with Kiribati and lose the use of the important space station, and third, ignore it. In all probability, the PRC will play it down, as it would not like to be seen to precipitate problems by arm-twisting small countries in the region but it can also not allow a tiny island nation to defy the PRC. This could well prove to be the first important test of the new 'fourth generation' leadership of China.

Sino-Indian Relations

In 1962, China went to war with India for two main reasons; first to remove the threat of Indian military intrusions into what it claimed as Chinese territory, and second, to teach India a lesson so that once, and for all times, India remained intimidated by China's military strength and political resolve. China also wanted India to realise that it could never aspire to join the same league as the bigger and more powerful neighbour. As seen in an earlier section, China, having signed a border treaty with Myanmar in 1960 and Pakistan in 1963, was now in a better position to coerce India into accepting a less favourable dispensation on the border. China also found it easier to intimidate India by covertly supporting insurgencies in India's North-east and openly threatening a two-front war in 1965 when Pakistan appeared to be losing the conflict that it had launched against India. In 1971, however, China protested but did not actually hold out a military threat when India and Pakistan once again went to war over East Pakistan. India had by then, done much to improve its defence preparedness and signed the Treaty of Friendship with the former USSR, while China was in the middle of the disastrous Cultural Revolution and had not as yet made much progress in its relations with the US. What is however more germane to the current discussion is the manner in which the PRC quickly changed tack. Perhaps in deference to Pakistan's wishes, the PRC was one of the last to recognise the newly liberated Bangladesh, but was to soon become one of the bigger donors and friends of the new regime after the assassination of Mujib-ur-Rehman in August 1975. For many years now, it has remained the most important source for arms and equipment to Bangladesh.

By 1976 India and the PRC had re-established diplomatic relations, and China had begun to reduce its support to the rebels in India's North-east, but it was not until 1988 that the relations showed any substantive movement forward. The then Indian prime minister, the Late Rajiv Gandhi's visit to Beijing was a significant landmark. Both sides have maintained contact since then and signed two treaties on maintaining tranquillity on the border in 1993 and 1997 respectively, but there has been little progress on ground. It will be remembered that during the brief Indo-Pakistani border conflict over Kargil in 1999 too, the Chinese had allegedly made some threatening moves along and according to some, even across,[14] the border but did not actually resort to military action. This was perhaps because on each of those occasions India seemed determined to continue fighting until its immediate political aims had been achieved but appeared disinterested in escalating the conflict to a level where the very existence of Pakistan was threatened.

The June 2003 visit of the Indian Prime Minister, AB Vajpayee was seen as successful since the two sides finally decided to name two senior political personages for further discussions on border delimitation. China extending its recognition to Sikkim as part of India by obliquely agreeing to opening a trade route through the Chumbi Valley, was also seen as a major achievement by some in India, but many others saw it only as a minor and a rather belated acceptance of a long established fact.[15] The accusations and counter-accusations that followed a newspaper report about Chinese intrusions across the Line of Actual Control, within days of the Indian prime minister's return to Delhi are a clear indication of how fragile the relationship really is. There is deep mistrust just below the surface and even a minor aberration opens the old wounds. While India has no doubt attempted to improve its relations with China while simultaneously building up its military strength, it is unlikely that it can actually bridge the power gap; in reality, the differential is likely to grow further. It is thus seen that while India must continue to build and if possible further deepen its relationship with the PRC, there are serious limitations to such an approach. China's strident reaction to (a) the May 1998 Indian nuclear tests and (b) the Indian defence minister's statement about China being India's 'Potential Threat Number One', is a clear indication of how sensitive the PRC is to Indian plain-speak. Trade, economic and cultural ties as also cooperation in addressing regional and global issues would no doubt be welcome and would perhaps be tried, but real trust would take long to build simply because of the PRC's nexus with Pakistan, and now, the defence treaty with Bangladesh. The only way China can win India's trust is by severely curtailing the military content of its relationship with India's neighbours and that it is unlikely to do.

Strategic Encirclement of India

The PRC's efforts to modernise its armed forces are aimed at making them effective instruments of coercion and intimidation. These efforts have been comprehensive and multidirectional. Military modernisation has been an integral part of building and enhancing China's Comprehensive National Strength (CNS) and hence the efforts are intimately linked to foreign policy initiatives and include access to technology and bases or transit facilities. China has slowly gained access to many parts of the region and it is now in the process of consolidating these gains. The process has been slow, even halting at times, but the final goal has been to improve the PRC's ability to influence events abroad. It is part of what Garver calls a 'plausible denial' approach. By helping these countries build their economic and commercial infrastructure such as airports, roads and ports, China can get a strategic access and yet deny that the assistance has any military content.[16]

China's relations with Pakistan, Myanmar, Bangladesh, Sri Lanka and Nepal are of greater military significance as these smaller countries lie within India's immediate security sphere and are vulnerable to economic and political manipulation. These relationships, therefore, impinge directly on India's security.

Myanmar

As already discussed in Chapter 2, China has been a long-time friend of Myanmar, a country plagued by internal strife. China not only curtailed the freedom of the Burmese Communists but also helped the Burmese government in its fight against the insurgencies. China then provided much economic and military aid so that over the last four decades, Myanmar, especially the present ruling Junta, has become increasingly indebted to the PRC. In return, the PRC has helped Myanmar in building a number of strategic roads and railways in the Northern part of the country. This has not only helped the Myanmar government in improving its control over the far-flung and hitherto inaccessible areas of the country, but has also facilitated trade and economic contact with the adjoining regions of China. In the long term, these routes would provide the PRC direct overland access to the Bay of Bengal and would help integrating the trade links between the two countries. These routes would also be militarily significant as the PRC can move large bodies of troops and heavy military equipment to the western shores of Myanmar from where these can be shipped abroad. As part of its effort to build infrastructure in distant areas, the PRC has undertaken to improve and upgrade facilities at a number of Myanmar ports. By the turn of the current decade, the PRC would have completed the ongoing construction of ostensibly civilian bases and infrastructure facilities from Sitwe, in the North, through Hainggyi, Kyaukpyu, Bassein, Mergui and Yangon, to the radar facilities on Coco Islands, and what Garver calls "the most militarily significant facility at Zadetkyi on Kyun island just off Myanmar's Southern-most territory on the Kra isthmus".[17] This would give the PLAN and the PLAAF many secure bases to operate in the Bay of Bengal while the 12,400 foot long runway near Mandalay would launch the lumbering, old, and nuclear capable H-6 (Tu-16) tanker/bomber force. Although Garver also sounds a note of caution about Myanmar allowing the Chinese Navy and Air Force the use of its bases in a future conflict, he seems clear that years of friendly exchanges and copious economic help to all these countries would prove very helpful in opening doors for Chinese activities.[18] In a future conflict scenario, the PRC will most probably call upon the Myanmar government to allow its armed forces the use of these facilities, failing which it could simply ignore and override Myanmar's neutrality.

Pakistan

Over the last four decades, the PRC has done much to bolster the security of this country. Soon after signing the border treaty in early 1963, the PRC gave it moral and political support during the 1965 War with India and was ready to open the second front. Lest it be forgotten, the PRC issued a deadline to India to unfreeze the assets of its Bank of China in Bombay and Calcutta and warned India of serious consequences should India fail to provide a 'speedy and clear-cut reply'. The PRC had some 60,000 soldiers across the border in Tibet to make good the threat. Garver opines that this Chinese threat perhaps expedited India's acceptance of the UN sponsored ceasefire on 22 September 1965, which the Pakistani leadership accepted only at the last moment. Soon after the end of the war and following the US embargo on shipment of arms to India and Pakistan, the PRC began a huge arms aid programme under which it gave Pakistan arms for two infantry divisions, 200 T-59 tanks and 120 MiG-19 fighters that were modern by the day's standards. The MiG-19 fighter-bombers were offensive weapons, and could deliver over 1000kg of armament; whereas the MiG-21 fighters that India had received from the USSR were purely for high-altitude air defence and carried only two air-to-air missiles. The Sino-Pakistan friendship has since then blossomed into the PRC giving it nuclear, missile and other technological assistance and modern arms. The large numbers of F-7 P (Modified MiG-21) fighters that Pakistan has received from China in the recent past are an important part of the Pakistan Air Force.

Further, the PRC built a heavy machinery complex near Taxila and later gave all out assistance to build nuclear reactors, the latest of which is a 300 MW reactor coming up near Lahore. China also constructed the, "Sino-Pakistan Friendship Highway" across the mountains in Pakistan occupied Kashmir that connects Kashgar in Western China's Xinjiang province. The highway has been open since 1978 and the two countries recently celebrated the 25th anniversary of its opening to third-country nationals. While logistically, it is no doubt difficult to transport heavy equipment over this mountainous road but in psychological terms, the highway has done much to boost the Pakistani morale. The PRC is now engaged in upgrading the port facilities at Gwadar on the Makran Coast that would serve the dual purpose of enhancing trade with Central Asia and more importantly providing access to Chinese Navy in peace and wartime. The first phase of the project is due to be completed by 2005 and if China's past record is any indication, it will be completed on time. Needless to add that Pakistan would be only too eager to allow the PRC the use of these facilities in any future conflict with India.

Bangladesh

As noted earlier, China's initiatives in befriending this newly liberated country began immediately after the assassination of Mujib-ur Rehman in August 1975. Since then the ties have grown so that today almost 90 per cent of the arms and equipment used by the Bangladeshi armed forces is of Chinese origin and includes fighter-bomber aircraft and naval surface vessels with considerable offensive capability. With the signing of a defence treaty in December 2002, the ties have been further strengthened and although both China and Bangladesh have stated that the treaty is not aimed at a third country it is evident that when needed, Bangladeshi ports, air bases and other infrastructure could be available for use by the PRC's armed forces.

Nepal

China also built a highway linking Kathmandu with Lhasa in Tibet and provided some small arms to this Himalayan Kingdom but vigorous Indian attempts to prevent deeper inroads into this country with which India has a special relationship, have restricted Chinese access to some extent.

Sri Lanka

This tiny island nation has also received considerable aid from the PRC especially after its relations deteriorated with India in the 1980s but of late India has managed to curb theses contacts to manageable levels.

Instruments of Coercion

The PRC has often forcefully denied that its help to India's neighbours has any anti-India element. The PRC insists that these are part of its normal bilateral relations and are not aimed at weakening India's security but there is no gainsaying the fact that China has, over the last four decades, systematically developed these ties and is now poised to reap the benefits of this heavy financial investment and diplomatic hard work. Due to its inherent economic and military weakness and a complete absence of aggressive spirit vis-à-vis China, India has had no option but to remain a mute witness to these dangerous developments in its close proximity. Let us now turn to a brief examination of the actual arms inventories that the PRC is likely to possess by 2010.

PLA Army

The PLA would be downsized to about 50 active divisions under combined or joint theatre commands in each of the seven MRs or military area commands (MAC), of China. It would be equipped with modern firepower, self-propelled artillery and state-of-the-art armour. It would be supported with modern satellite based communications and C4ISR and computerised logistics to enhance its fighting ability. A variety of Russian origin, air defence missiles would give it considerable protection and a whole host of foreign and indigenously produced helicopters will add to its combat edge. Increasingly larger numbers of its formations would be capable of rapid reaction and the huge civilian commercial air fleet of the PRC would permit large-scale strategic mobility. The PLA would, thus, be in a position to ward off any major threat to China's continental border areas and be prepared to move into a distant theatre by sea or air.

PLA Navy

With the induction of two Sovremenney class destroyers in recent years, the PLAN offensive capability is already high. Two more of these modern destroyers would join the PLAN by the end of the decade and give it a near blue-water capability. More Jinagwe class modern frigates, Kilo class submarines, improved Song class indigenous submarines along with state-of-the-art-missiles, torpedoes and all round air defence capability would remarkably transform the Chinese

Navy. The addition of more J-8II fighters and JH-7 fighter-bombers to the aviation arm of the navy would further reduce its vulnerability on the open seas and add to its survivability even without an aircraft carrier. The PLAN would also induct a sizeable number of more capable amphibious assault boats and equipment, thus, enhancing the PRC's capability to deal with any threats to its vast island territories in the South China Sea. By 2010, the operational reach of the PLAN would also have far expanded and the PRC would be in a better position to undertake sea denial and sea control operations although at present, the PLAN does not expressly mention these tasks as part of its mission.

PLA Air Force

The PLAAF would be the sword-arm of the PRC's power projection capability and hence needs to be watched most carefully. For many years, it possessed a huge inventory of some 5,000 combat aircraft. A majority of these were obsolete and, hence, useless for any meaningful air operations. Over the last decade, by shedding most of its old aircraft, the PLAAF has become more manageable and combat-worthy. After absorbing the lessons of the 1991 Gulf War, the PLAAF has made concerted efforts to modernise its equipment, doctrine and offensive strategies. China's efforts to produce a large variety of fighters, transports and helicopters have also succeeded with copious assistance from Russia, Israel and many other European countries. During the heyday of its friendship with the US, China received major injections of high technology such as high-speed computers and rare insights into US Air Force tactics, training, and doctrine. The professionalising of the PLA also helped speed up the process of its modernisation. More and more senior cadres of the Party and officers in the armed forces saw the wisdom in giving the PLAAF a freer hand in shaping its force structure. The upshot of these favourable developments was the deep national commitment to air power.

modernisation, rapid induction of modern aircraft such as the Su-27/30, Il-76 and missiles and a fillip to indigenous design and production. After much effort and generous funding, the Chinese aviation industry is now in a position to roll out large numbers of FC-1 (Super-7) and J-10 fighters powered with the Russian RD-93 and AL-31 F engines respectively. In addition, these fighters would be fitted with modern airborne radars, avionics and other facilities such as mid-air refuelling, Self-protection jammers and chaff/flare dispensers as part of the Electronic Warfare suits and air-to-air and air-to ground missiles and PGMs. Most importantly, the PLAAF would be in a position to field a large number of these aircraft. Table 9.1 details the changes in the PLAAF inventory.

Table 9.1

Aircraft Type	**1985-86**	**1995-96**	**2000-01**	**2002-03**	**2010 Forecast**
Fighters					
J-6 B/D/E	3,000	3,000	1,500	500	NIL
Q-5	500	400	300	300	NIL
J-7	200	500	400	654	1,000+
J-8 II	30	100	250	204	500
SU-27	NIL	22+04	65	70+20	100
SU-30	NIL	NIL	FEW	38*+38	200
FC-1	NIL	NIL	NIL	FEW $	100+
J-10	NIL	NIL	NIL	FEW	300+
JH-7	NIL	NIL	NIL	50 ^	100
J-11#	NIL	NIL	FEW	50?	200
TOTAL	3,730	4,000	2,515	1,824	**2,500**

Source: *IISS Military Balance* (Oxford University Press), for the years shown in the table, Figures for 2010 are own estimates.

Note: Figures include all variants and marks.

* Another batch of 38 was delivered in early 2003.

$ Production formally commenced in early 2003, 150 ordered by Pakistan, the partner in the joint programme.

Indigenously built Su-27 or Su-30.

^ JH-7s were to be built for the PLA Navy's aviation wing, actual figures and whether these would come to the PLAAF is not known.

As shown in the table, by 2010, the PLAAF would once again possess some 2,500 fighters but the most remarkable aspect of this inventory is that all of these would be relatively new and modern aircraft capable of maintaining high sortie/utilisation rates. In other words, the PLAAF would be in a position to launch a bigger number of offensive long-range precision strike and air defence missions over larger distances and for longer durations. Given the improvements in the PLAAF logistics system, these aircraft would be provided full and timely spares-support even when these are launched from forward airfields that normally do not base them.

Second Artillery

As discussed earlier, this important branch of the PLA would also field modern, more accurate and a larger inventory of nuclear and conventional missiles. According to one estimate, the number of DF-15 missiles alone would be nearer 650. These and other missiles and cruise missiles would be GPS guided and launched against targets for which more accurate data/coordinates are obtained through the PRC's reconnaissance satellites.

Summary

To summarise, by the end of the current decade the PRC's armed forces would include the following:-

(a) A downsized, smart, digitised, modern army with ultra modern C4ISR, and Special Operations Forces capable of rapid reaction and strategic deployment over considerable distances.

(b) A modern navy with near blue-water capability that will dominate the South China Sea, Western Pacific and the approaches to the Indian Ocean Region (IOR). Even without an aircraft carrier, the PLAN would be able to support its operations over long distances from its bases in the South China Sea Island Territories, Myanmar, Bangladesh and from the port facilities at Gwadar on Pakistan's Makran coast so that the PLAN can effectively control the approaches to the Persian Gulf.

(c) A long-range precision strike capability based largely on the PLAAF's Su-30, J-10 and FC-1 fleets and a balanced air arm without a peer in the region.

(d) A space and missile capability that would include accurate and even Multiple Re-entry Vehicle based nuclear weapons supported by a whole host of conventional missiles and SRBMs, navigation and reconnaissance satellite launching capability that ensures full spectrum dominance.

Possible Future Scenarios

Writing in an Indian weekly, Humphrey Hawksley, the well-known commentator for the BBC and celebrated author of the book *Dragonfire*, considers four possibilities that might trigger a conflict that could further escalate to regional level and in each of them the main players are China and India. He says, "Should Tibetan separatists abandon their official policy of non-violence and take up weapons again, India will come into conflict with China." He adds, "Should Pakistan or India cross the line of Control again; should Pakistan feel its existence is being threatened; should China choose to weaken India by increasing its help to Pakistan over Kashmir; should China choose to test India's naval power by pushing further into the Indian Ocean; should China demand that India end its sanctuary for Tibetan exiles- should any of these variables happen, then these two great emerging nations will find themselves in conflict".[19] Even nearly three years later, the scenarios remain equally plausible.

Of these possibilities, a conflict with Pakistan and a sudden and unexpected uprising by the Tibetan activists in India or even inside Tibet appear the more dangerous precisely because these are more likely. Despite many sincere efforts by a very enlightened and realist Indian leadership to lower the temperatures on the subcontinent, Pakistan has refused to see reason. There is little sincerity in the commitments of its leadership. Their utterances are often bellicose and offensive. Given the fragile opposition in that country, there is little chance that the military leadership would ever pay heed to the cautious voices of its remarkably courageous media. Pakistan continues to support cross-border terrorism even when it is clear that it is a self-defeating pernicious policy. It should by now be evident to Pakistani policy makers that the world at large does not believe its double-speak and does not approve of its nuclear brinkmanship. So strong and visceral is the Pakistani establishment's hatred for India that it appears determined to confront India, whatever the cost to its own well-being may be. Its dependence on the PRC to derive a false sense of equality with India and the latter's reluctance to openly condemn Pakistan's behaviour because of its short-term usefulness do not quite mesh with China's friendly declarations towards India and hence without a radical change in its approach to the sub-continental security issues, India would have no option but to treat China as an extant threat. As in the past, Pakistan may well up the ante if China's covert support is assured. China might not have openly supported Pakistan during the 1999 Kargil conflict, but it is hard to believe that it was not consulted prior to Pakistan launching its ill-fated and unwise infiltration across the LOC in the Indian state of Jammu and Kashmir.

The Tibetan problem is also capable of spiralling out of control once the benign influence of the Dalai Lama disappears. It is quite likely that the PRC would automatically blame India for any uprising that might occur unbeknown to it. In such an eventuality, the Indian leadership might find it extremely difficult to use its influence to subdue the feelings of Tibetan upstarts supported by unknown entities in the West. The PRC could well misconstrue this inactivity on India's part as tacit support to Tibetan uprising and decide to retaliate with force.

The PRC's military action against India might begin with a warning to India of dire consequences for its fictitious, imaginary and non-existent support to the Tibetan rebels. China could launch a number of 'low-earth-orbit' reconnaissance satellites; move its sizeable SRBM assets menacingly close to India's borders in Sichuan, Yunnan or even Tibet. It could move some one hundred planeloads of its elite Rapid Reaction Troops to air bases in close proximity of the border in Tibet and increase the activity of PLAAF reconnaissance aircraft, and Unmanned Aerial Vehicle (UAVs) in full view of Indian troops. China could deploy its long-range fighter-bomber fleet of Su-30s in Sichuan, Myanmar or even Pakistan and Bangladesh; move its H-6/Il-76 tankers to Myanmar; deploy its submarines and destroyers in the Bay of Bengal and Arabian Sea and then issue an ultimatum to India with a short deadline. To substantiate the threat, the PRC's ambassador in Delhi could provide the Indian government with high-resolution photographs of India's Agni nuclear missile storage sites complete with launch vehicles, foreclosing all Indian options. The firing of a warning salvo of a few DF-15 missiles across into Assam could well begin an unmanageable refugee exodus from the distant northeast. The United Liberation Front of Assam (ULFA) and Manipuri insurgents could well take advantage of the situation and strike behind own lines.

A calibrated bomber raid on Calcutta and/or Guwahati airports could well force the Indian leadership to rethink its options. Although communications are now much better, let it not be forgotten that the local government functionaries fled the scene at the first signs of a possible Chinese advance into the planes in 1962 and the Central Government at Delhi appeared to abandon the hapless people of the border regions. What is to stop this from happening again? What if any course of action would then be open to India? Would India then have any option but to quietly accept Chinese terms?

These scenarios are the more plausible for many reasons. For the foreseeable future, China is unlikely to precipitate the situation across the Taiwan Straits. It would also not like to confront the US when its economy is so closely linked to it. The smaller players in the region are unlikely to oppose a mighty and obviously aggressive China. In sum, China would not face any threat either from the north or from the south and may well find itself free to administer another lesson to an imaginary 'recalcitrant' India.

Conclusion

Does it then mean that India is doomed to accept a subsidiary role in the region and live with what dignity it might be able to muster from a skewed power equation with China? The answer is emphatically in the negative. Provided India girds its loins and embarks on a sincere and concerted campaign to enhance its overall power and international leverages, it is still possible to arrest the adverse power differential that India has allowed to build with China. India would however have to get its act together. In the post-1998, nuclear tests era India's foreign policy has been significantly proactive and innovative.

First, the Indian decision making elite must appreciate that India is the only country in the region that has accepted changed status quos in the past to the detriment of its security. This was due mainly to its overall military weakness and an almost pacific approach to national security. As K. Subramaniam has often said, "Indians have often needlessly politicised security." A mere promise of a hiked defence budget does not automatically translate into increased security and military power. Second, India must seriously review its force structure and revamp the armed forces in a manner that their deterrent content is enhanced. It must also develop a national security doctrine that leaves little doubt about India's political resolve. India must simultaneously build its capacity to deter coercion by selectively developing its nuclear and conventional offensive capabilities. Acquisition of new 'aircraft and weapons systems' alone does not add to national political resolve. The armed forces too need to play a more proactive role. It was reported that during the face off with Pakistan in 2002, the armed forces were reluctant to go to war because they feared that they did not enjoy the minimum necessary numerical superiority and hence could not assure a favourable outcome.[20] Such reluctance, if true, would have seriously diminished the credibility of India's security apparatus. India needs to develop what I would call 'strategic poise' that helps demonstrate its national resolve.

Third, India must invest at least one per cent of its defence R&D budget on strategic studies. Without a dedicated team of multidisciplinary scholars and China watchers including persons well versed in RMA technologies, it would be difficult if not altogether impossible to keep track of China's game-plan and this might reduce reaction time. Fourth, India must take a deeper interest in Central and Southeast Asian affairs. Sending teachers, professors, and professionals like doctors and engineers and increasingly large numbers of popular film stars may pay disproportionately high dividends. Fifth, India should continue to build its ties with the countries of Central Asia as it has recently done with Tajikistan. Finally, India needs to vastly enhance and reenergise its military diplomacy by significantly increasing the number of service officers that it sends on foreign training and diplomatic appointments. As has been brought out earlier, the PRC has, over the last decade increased its military contact so dramatically that today this number has crossed 102 countries as against a mere 35 that India maintains.

Finally, the PRC today has a set of young, dynamic, well-informed, well-travelled, tech-savvy, forward-looking leaders who are determined to sustain the high rates of progress in every department. They are conscious of the role of the Communist Party in ensuring a relatively peaceful transition to modernity. "This is a group of men who believe in authoritarian rule as a precondition for pushing their country through the turbulent passage of modernisation. They believe in stopping dissent against Communist rule before it gets started and in deterring crime by widespread use of the death penalty. They believe that their government has been more than generous to the residents of Taiwan, Tibet and Xinjiang and have little sympathy for the grievances of the people in these places. Although they share many Western economic values, they share fewer Western moral values." "The Party, the Government and the Military are separate on paper, they are inseparable in reality."[21] Indians need to understand this reality.

The PRC has a penchant for spectacular demonstrations of its power. As the late Air Chief Marshal PC Lal had stated, "In the 1950s, when an Indian military delegation visited China, the PLA staged a show. The visiting dignitaries were taken to a barren area with little vegetation. Suddenly a huge number of guerrilla fighters apparently materialised from nowhere, fired guns and rockets and disappeared as quickly into underground bunkers as they had emerged." This impressed the Indians and the PLA generals were happy that China's offensive potential was amply demonstrated. In 1999, the PRC said that it would put a

Chinese in space in three years and it successfully did so on 15 October, 2003. In 1995 and 1996, the PLA fired missiles across the Taiwan Straits in a show of strength. It is quite likely that with the staging of the 2008 Olympics, the PRC would once again demonstrate to the whole world its national power, but this time in a benign manner. It is also likely, that China will indeed build its own space station in the next three to five years. Militarily these demonstrations might not appear too significant but ask a common layperson in the region and he would readily acknowledge the PRC's power.

At first glance, these estimates might appear somewhat alarmist. It is understandable that those engaged in routine foreign policy dialogue with the PRC would perhaps not like to voice these concerns so explicitly. Given a marked sophistication in the PRC's conduct of foreign policy in the recent past and its attempts to improve relations with India, the reaction of Chinese scholars and analysts to these scenarios may well be angry. There is the danger of these forecasts being termed as wildly exaggerated and fantastic even by some Indian observers. It should, however, be remembered that no one can really fathom a country's intentions. This study has made an honest attempt to forecast the future capabilities of an emerging major power based on past events and has considered only the actual ground realities. It is not as if China would resort to military force at the very first instance but then it cannot readily explain why it has so painstakingly built and optimised its military capabilities for a contingency in the region especially in South Asia. As noted earlier, China would not countenance similar foreign policy initiative by India in its neighbourhood, say, in Mongolia, Vietnam or even Japan. It is time the Indian military started to seriously address the real challenge to India's security and mentally disengaged itself from the insignificant, albeit, costly threat from Pakistan.

Notes

1 *Time,* 2 May 1994, pp 23-24

2 Wu Xinbo, "China: Security Practice of a Modernizing and Ascending Power", in Muthiah Algappa, *Asian Security Practice* (Stanford University Press: Stanford, California, 1998), notes on p. 709.

3 Andrew G. Nathan and Bruce Gilley, *China's New Rulers: The Secret Files* (Granta Books: London, 2003), p. 5.

4 Ibid. p. 49.

5 See China's Defence White Paper 2002 and its 'New Security Concept' available on the internet at<http://china.gov>

6 *Indian Express,* "India's Military Base in Central Asia".

7 Vijay Sakhuja, "China's Maritime Power", IDSA paper.

8 Amitav Acharya, "China and Southeast Asia: Security Aspects", paper presented at IDSA Annual Asian Security Conference, New Delhi, January 2003.

9 I am grateful to Lieutenant General Satish Nambiar (Retd), Director USI of India, for bringing this point to my attention.

10 Valerie Niquet, in the *USI Journal.*

11 Wu Xinbo, n.2, , p.151.

12 Ibid. pp. 148-149

13 *People's Daily* Online, Internet Edition, 08 November 2003, and Free Intelligence Summary issued on 07 November 2003 by <http://www.stratfor.com>

14 Although, officially the Indian Government did not complain about Chinese military moves during the Kargil episode of May-July 1999, there were strong rumours in Delhi of such Chinese intrusions; at least one analyst has written about these in a defence journal. See for example, Brigadier Vijay Nair in *Defence Technology,* March 2003.

15 Various newspaper reports in the Indian press in June-July 2003.

16 James Garver, *Protracted Contest: Sino-Indian Rivalry in the Twentieth Century* (Oxford: Oxford University Press, 2001), Indian Edition, p.380.

17 Ibid. pp. 294-295.

18 Ibid. p.295.

19 Humphrey Hawksley, "Future Wars Flashpoints: Will there be a Third World War?", in *The Week*, 31 December 2000, p. 99.

20 Manoj Joshi, in IDSA's *Strategic Digest* of February 2003.

21 Andrew G. Nathan and Bruce Gilly, n. 3, pp.4-6.

10

Afterword

Peace and Stability in South Asia

This study is about the China's power projection capabilities. As we have seen, a nation's power is not simply the number and types of ships, submarines and aircraft it possesses, but the sum total of its exertions in several areas. Diplomacy, politics, international relations, economic development, strategic weapons, military doctrine and its long term vision of the world and most importantly the role it wishes to play on the global stage are some of the vital factors that contribute to the PRC's sustained efforts to enhance its Comprehensive National Power (CNP). China's strategic vision began to evolve immediately after its liberation under Mao and its leadership has diligently and painstakingly worked to transform that vision into reality.

A more important aspect of this study, however, is the implication of China's burgeoning national power on India's security and here what matters most is how the PRC has expertly employed selective economic, nuclear and military assistance to some countries in South Asia to make up for the shortfalls in technology, equipment, weapon systems and long-reach naval capabilities. As discussed in these pages, it is China's special relationship with Pakistan, Bangladesh and Myanmar that poses the gravest challenge to India's security. China's problems with the restive Tibetan population and the long outstanding border dispute only exacerbate the adverse relationship.

Throughout this study, a timeline of the year 2010 has been used to forecast the actual capabilities that the PRC may acquire by that time. That timeline, however, is six years away and China's CNP could well reach the critical mass before 2010. This would largely depend on how it uses the present and future opportunities and how other major powers deal with a powerful China that is determined to change the rules of the game. To get an insight into this, we need to briefly take a closer look at what the Chinese call 'SHI' or the 'current strategic configuration of power'. The Chinese strategic analysts, no doubt, are continuously studying it.

We are now in a post-Post-Cold War era. The Cold-War calculations are no longer relevant to forecast future events. Since the cataclysmic terrorist strikes of September the 11th 2001, the US has been preoccupied with fighting global terrorism and does not wish to be distracted from this priority. The invasion of Iraq was expected to strengthen America's ability to route out the twin threats of terrorism and WMD. The Iraqi problem in the aftermath of the war, however, has only become even more complicated, both militarily and politically. Iraq is far from peaceful even after the dramatic capture of Saddam Hussein. Although the war has given the US a foothold in the heart of turbulent West Asia, the spectre of instability in Saudi Arabia and WMD threats from Syria loom large on the horizon. The Al Qaeda is still active and even if they have not been able to stage a spectacular attack since 9/11, frequent terror alerts demonstrate their ability to frighten the industrialised countries with destruction and disruption. Despite the much publicised 'road map', the Israeli-Palestinian conflict goes on unabated without any let up in suicide bombings.

Afghanistan too continues to simmer with no sign of the Al Qaeda or the Taliban leadership falling in the American dragnet. It can at best stabilise as a loose arrangement where local warlords exercise sovereign powers while the Kabul government's writ is restricted to its immediate environs. The US has no option but to mount a major ground operation to flush out the remnants of the Taliban and Al Qaeda leadership from the border areas of Afghanistan and Pakistan. The opposition elements in Pakistan have, however, warned the US against sending its army into its Tribal Areas. This means that Osama Bin Laden can peacefully live out his old age in the company of Mullah Omar.

Other US responsibilities and concerns such as North Korea and Taiwan, the former due to its intransigence on the nuclear weapons issue and the latter because of its incremental demands for independence, together do not permit the US to thin out its military presence in the region. The US forces are thus stretched and restrict the possible application of military force in any future eventuality. According to one analyst, America's first unilateral pre-emptive strike may well be its last.

As if this was not enough, the recent revelations of its 'ally of allies', Pakistan, giving nuclear know-how and material and possibly even weapon designs to Libya, Iran and North Korea, all premier members of the 'axis of evil', is a major problem that would give many more sleepless nights to the US policy makers. Its immediate 'official' reaction shows that global terrorism has now overtaken the threat of nuclear non-proliferation. By giving the 'presidential pardon' to Dr. Abdul Qadir Khan, the self confessed villain of the sordid drama, Pakistani authorities may think that the matter is closed but even its own people refuse to believe the official version. The US has somewhat belatedly demanded access to the findings of the probe. The US media is amazed and appalled by the response of the Bush administration and is asking it to keep a close watch on Pakistan's activities. India has rightly suggested that the transfer of nuclear technology and material to other countries cannot be dealt with as an internal matter of Pakistan and should be scrutinised and discussed at international fora. The Bush regime went to war against Iraq for suspicion that it was in the process of developing its WMD programme. No evidence of Iraq possessing WMD has surfaced so far. It is indeed strange that Pakistan is let off even when it has blatantly proliferated to America's enemies. Pakistan has proved the redundancy and the inherent weaknesses of the NPT. According to some observers, however, the US is increasing pressure on Musharraf to take urgent military action to capture Osama Bin Laden or quietly accept the US army's entry into Pakistan's North-West areas. There are also reports of the US guarding the Pakistani nuclear weapons lest they fell in the hands of Jihadi elements.

Iran faces further political conflict between the conservatives and moderates and it is not clear if the latter would be allowed to take a more conciliatory approach towards the US. The recent disclosures about Pakistan's complicity in providing it with nuclear technology may further complicate the situation. For the present, Iran is cooperating with the US in stabilising Iraq but this may well take considerable US effort and resources. There is no doubt that China would take advantage of this uncertain situation. This being the election year in the US, there is every possibility of populism overtaking sound strategic decision making.

Russia too continues to remain bogged down in Chechnya with an increase in the frequency and intensity of terrorist strikes in Moscow and other cities. Although it has improved its economy, it continues to be a weak power unable to influence events in Central Asia, and the North East Asian region; once again giving China a clear run. Russia's far-east also remains vulnerable to Chinese trade, economic expansion and demographic incursions.

Although Japan has finally sent a few hundred troops to Iraq, it is far from clear if its non-pacifist elements would actually win the debate over Japan's larger strategic role in the region. In all probability, Japan would eschew use of force and counter the rapidly increasing Chinese power by further strengthening its economic ties with it. The PRC, however, counts Japan as an extant threat and criticises its defence spending.

Three recent developments in the region, the proposed Indo-Pakistan peace process, revelations of Pakistan's role in nuclear proliferation and India's strong economic resurgence, and finally the possibility of the BJP returning to power with an even bigger majority at the head of the multi-party coalition, may deeply transform the region. Prime Minister AB Vajpayee had announced a peace initiative on 18 April 2003. Although it has taken some ten months of coaxing and cajoling before Pakistan finally decided to clamp down on the terrorist networks, winds of change are finally blowing over the region.

Terrorism may lose popular support in J&K and Pakistan if the peace process shows signs of success. This time around Pakistan seems a little more earnest but how long that sincerity lasts would depend on Pakistan's internal situation and the US determination to maintain pressure on Musharraf. What is in it for Musharraf who had said that even if the Kashmir issue were resolved animosities would remain between Pakistan and India? Many Pakistani Generals have often declared that Pakistan would not rest until the crescent and star flew from the ramparts of the historic Red Fort in Delhi. Pakistan has wanted to destabilise India as its very success as a secular democracy is antithetical to Pakistan's ideology. India's economic success is a threat to its army dominated establishment. Pakistan's aim is not simply to wrest Kashmir. Why then has it agreed to peace talks? Some say the US has done much arm-twisting. Condoleezza Rice, Richard Armitage, Colin Powell and President George W. Bush himself, have persuaded Musharraf. Some also say that Musharraf cannot survive new threats from Islamic fundamentalists in Pakistan without mending fences with India.

The potential benefits of a peaceful subcontinent are undoubtedly enormous. Not only would terrorism against India reduce but even at the global level it may lose its appeal. The moment Pakistan accepts that a quiet and peaceful LOC is in its interest, it should be quite easy to reduce Indian army's presence in J&K. Free two-way movement of ordinary people across the IB and the LOC would certainly calm tempers. India could export tons of mangoes, betel nut, leaf, tea and a whole host of engineering goods and computer software to Pakistan. Exchange visits of cricket and hockey teams, musicians, TV and screen personalities can build huge goodwill and lower the temperature and perhaps change the perception of ordinary Pakistani and Indian citizens towards each other. Pakistan could also benefit from Indian expertise in medical, engineering and software education. Pakistani students could benefit through higher education in India. Such long term contacts may help build trust at several levels of interaction. This is, however, possible only if Pakistan agrees to develop the dialogue process without insisting on first resolving the 'core' issue of Kashmir.

According to the latest estimates, Indian economy has performed much beyond expectations despite global recession and disruptions caused by terrorism, SARS, Bird Flu and Iraq War. The 2003 Third Quarter results showed the economy growing at an unexpected 08.24 per cent and good monsoons and a new confidence in India's intrinsic strength are said to be favourable for further Foreign Direct Investment (FDI). Some analysts, however, maintain that increased confidence in India's internal strength would unleash considerable amounts of funds within the country. The amazing and sustained rise of the Indian stock market 'Sensex' beyond the 6000 mark is evidence that the economic prospects are good. Improving trade relations and possible transfer of high-technology from the US may further catalyse the economic growth. The Pakistani media taking note of these developments has asked its leadership to concentrate on wealth generation rather than on continuing hostilities with India.

The revelations of Pakistan's role in nuclear proliferation are bound to eventually indict China for its pioneering role in this murky business. Notwithstanding the 'pardon' to Dr. Abdul Qadir Khan, and the relatively restrained reaction of the US, Pakistan is unlikely to remain unscathed. Some reports in the media allude to an even deeper role of the US in safeguarding Pakistan's nuclear arsenal and by extension lowering the chance of Pakistan resorting to nuclear sabre rattling against India. If Musharraf's statements during the 'India Today Conclave' of 13 March are to be believed, Pakistan seems to be adamant on first resolving the 'Kashmir' issue, again putting a question mark on the prospects of peace.

Last but not least, political developments in India show that the probability of a hung parliament is indeed very high. While elements inimical to India would try their utmost to prevent a two-third majority to any single party, chances are that the winner may still lead a stronger and more cohesive coalition. China and Pakistan have not forgotten that it was the BJP that had finally taken the difficult decision to stage nuclear tests in the face of strong external pressures and the consequent sanctions. They would naturally not like the idea of this party further consolidating its position with the Indian electorate. This is because the Left of Centre opposition parties have traditionally taken a softer line towards Indian security.

In view of the foregoing, it is more than likely that the PRC may not truly be happy at the prospect of peace between India and Pakistan but may outwardly support it. China cannot afford to throw away the gains of its long cherished relationship with Pakistan as its very useful and reliable ally in South Asia.

Pakistan is after all China's Israel. Indo-Pakistan friendship would not only reduce the latter's dependence on China but would also have a positive effect on India's other smaller neighbours such as Bangladesh, Nepal, Sri Lanka and Myanmar. These countries may well find that improving relations with India is more beneficial to them than playing the China Card. That would mean that China's long-term investments in these countries would begin to pay diminishing returns and its overall position in South Asia would weaken. The question is whether the PRC would actually permit such a situation to develop when it is the biggest loser.

In November 2004, at the 10th summit of the Association of Southeast Asian Nations (ASEAN), Prime Minister Manmohan Singh stated that the Indian government attached great importance to the development of bilateral relations with China and was keen to seek new ways for mutually-beneficial cooperation. The Chinese Premier Wen Jiabao voiced China's willingness to make joint efforts with India to expand and deepen cooperation in various fields. He added that China hoped to handle ties with India "from a strategic and overall point of view." India has also suggested to China "mutual accommodation" based on "ground realities" to resolve the boundary dispute. Wen has acknowledged that the boundary issue was a complex one requiring "time to resolve it".

China has also expressed willingness to further strengthen bilateral military ties with New Delhi. The Deputy Chief of Staff of Chinese People's Liberation Army (PLA), General Xiong Guangkai, described the visit of the Indian Chief of Army Staff General N C Vij to China as "a big event for the militaries of our two countries".

Wen's visit to India in March 2005 could "send a positive signal throughout the world," of Sino-Indian relations. At present, there are no clear indicators as to the shape of things to come, but if India plays its cards correctly and actually achieves lasting peace in the sub-continent its future economic and strategic growth would more or less be assured. China's efforts to enhance its usable power would no doubt continue but India would then be in a position to at least reduce the power differential to manageable levels.

Bibliography

Primary Sources

Government Reports

China's Defence White Papers- December 2002, October 2000, 1998.

PRC's New Security Concept.

Indian Ministry of Defence: Annual Reports.

Indian Ministry of Finance: Annual Reports.

Indian Ministry of External Affairs: Annual Reports.

Secondary Sources

Books

Alagappa, Muthiah (ed), *Asian Security Practice* (Stanford: Stanford University Press, 1998).

Allen, Kenneth W, Glenn Krumel & Jonathan D Pollack, *China's Air Force Enters the 21st Century* (Project Air Force, Santa Monica, RAND, 1994).

Allen, Kenneth W & Eric A McVadon, *China's Foreign Military Relations* (Washington DC: Henry L. Stimson Center, 1999).

Bajwa, JS (Brig), *Modernisation of the PLA* (New Delhi: Lancer Publishers, 2002).

Bertsch, Gary K, Seema Gahlaut, and Anupam Srivastava, (eds), *Engaging India: US Strategic Relations with the World's Largest Democracy* (New York & London: Routledge, 1999).

Dalvi, JP (Brig), *Himalayan Blunder. The Curtain-Raiser to the Sino-Indian War of 1962* (Bombay: Thacker, 1969).

Das, Durga, *India from Curzon to Nehru and After* (New Delhi: Rupa Books, 1973).

Evans, Richard, *Deng Xiaoping & the Making of Modern China* (New York: Penguin, 1993).

Garver, John W, *Protracted Contest: Sino-Indian Rivalry in the Twentieth Century* (Seattle: University of Washington Press, 2001).

Ji, You, *The Armed Forces of China* (London & New York: I. B. Tauris Publishers, 1999).

Joshi, Akshay, *Information Age and India* (New Delhi: Knowledge World and IDSA, 2001).

Kaul, BM, *The Untold Story* (Bombay: Allied, 1967).

Khalilzad, Zalmay M, et al, *The United States and a Rising China* (Project Air Force, Santa Monica: RAND, 1999).

Kondapally, Srikanth, *The People's Liberation Army: Evolving Dynamics* (New Delhi: IDSA 1996).

Lewis, John Wilson & Xue Litai, *China's Strategic Sea Power: The Politics of Force Modernisation in the Nuclear Age* (Stanford, California: Stanford University Press, 1994).

——————, *Uncertain Partners* (Stanford, CA: Stanford University Press, 1993).

——————, *China Builds the Bomb* (Stanford, CA: Stanford University Press, 1986).

Lal, PC, ACM, *My Years with the IAF* (New Delhi: Lancer Publishers, 1983).

Lilly, James R & David Shambaugh (eds), *China's Military Faces the Future* (Washington DC: American Enterprise Institute for Policy Research (AEI) 1999).

Mansingh, Lalit, et al (eds), *Indian Foreign Policy: Agenda for the 21st Century*, Vol.2 (New Delhi: Konark Publishers, 1998).

Mansingh, Surjit (ed), *Indian and Chinese Foreign Policies in Comparative Perspective* (New Delhi: Radiant Publishers, 1998).

Maxwell, Neville, *India's China War* (Bombay: Jaico Publishing House, 1970).

Mulvenon, James C & Richard D Yang, *The People's Liberation Army in the Information Age* (Project Air Force, Santa Monica, RAND, 1999).

Nathan, Andrew G and Bruce Gilly, *China's New Rulers: The Secret Files* (London: Granta Books, 2003).

Palit, DK (Maj Gen), *War in the High Himalayas: Indian Army's Debacle in 1962* (New Delhi: Lancer Publications, 1993).

Pillsbury, Michael (ed), *China's Views of Future Warfare* (Washington DC: National Defence University Press, 1997).

Ranganathan, CV & Vinod Khanna, *India and China – The Way Ahead after Mao's India War* (New Delhi: Har Anand Publications, 2000).

Segal, Gerald, *Defending China* (New York: Oxford University Press, 1985).

Shambaugh, David & Richard D Yang, *China's Military in Transition* (Oxford: Clarendon Press, 1997).

Singh, Jasjit (ed), *Asian Strategic Review* (New Delhi: IDSA, 1999).

——————— (ed), *Kargil: The Fourth Indo-Pak War, 1999* (New Delhi: Knowledge World, 1999).

Strachey, John, *The End of Empire* (London: Victor Gollancz Ltd, 1959).

Swain, Michael & Ashley Tellis, *Interpreting China's Grand Strategy: Past, Present and Future* (Project Air Force, Santa Monica: RAND, 2000).

Whiting, Allan, *The Chinese Calculus of Deterrence: India and Indo-China* (Ann Arbor: University of Michigan Press, 1975).

Zweig, David, *Internationalizing China: Domestic Interests and Global Linkages* (Ithaca & London: Cornell University Press, 2002).

Articles

Acharya, Alka, "Prelude to the Sino-Indian War: Aspects of Decision Making Processes During 1959-62," *China Report,* Vol. 32, No. 4 (Oct-Dec 1996), pp.363-393.

Bakshi, Jyotsna, "Russia-China Boundary Agreement: Relevance for India", *Strategic Analysis,* Vol. XXIV, No. 10 (January 2001), pp. 1833-1862.

Bristow, Damon, "China's PLA Modernisation and Regional Concerns," *Asian Defence Journal* (7/8 2000), pp.4-7.

Feigenbaum, Evan A, "Who's Behind China's High-Technology Revolution?", *International Security,* Vol. 24, No.1 (Summer 1999), pp. 104-136.

Foot, Rosemary, "Chinese-Indian Relations and the Process of Building Confidence: Implications for the Asia-Pacific," *Pacific Review,* Vol. 9, No.1 (1996), pp. 58-76.

Hawksley, Humphrey, "Future Wars and Flashpoints: Will There be a Third World War?", *The Week,* 31 December 2000.

Kristensen, Hans M, "China's Nuclear Assets," *SIPRI Annual Report 2002,* Appendix 6A.

Lewis, John Wilson & Xue Litai, "China's Search for a Modern Air Force," *International Security*, Vol. 24, No. 4 (Spring 1999), pp. 71-92.

Whiting, Allen S, "China's Use of Force," *International Security*, Vol. 26, No. 2 (Fall 2001), pp. 103-131.

Prasannarajan, S, "Who's Afraid of China?", *India Today*, 30 June 2003, pp. 32-34.

Yihong, Zhang, "China's Rising Forces," *Janes International Defence Review*, (August 2002), pp. 36-39.

Zuberi, Matin, "Soviet and American Assistance and the Pace of China's Nuclear Tests," *Strategic Analysis*, Vol. XXIV, No. 7, pp. 1247-1266.

Papers

Acharya, Amitav, *China and Southeast Asia: Security Aspects*, Presented at the IDSA Asian Security Conference, New Delhi, January 2003.

Bakshi, Jyotsna, *Post-Cold War Russia-China Relations: Relevance for India*, New Delhi, IDSA Fellows Seminar, 01 September 2001.

Chen, Vincent Wen-Hsien, *US Factor in the Cross-Strait Relations Between China and Taiwan*, New Delhi, IDSA Fellows Seminar, 26 October 2002.

Dutta, Sujit, *China's Territorial Sovereignty: Changing Notions of Frontier and Periphery*, New Delhi, IDSA Fellows Seminar, 11 November 2000.

Fisher, Richard D, *The PRC's Third Zhuhai Air Show*, 5-12November 2000, Lecture delivered at the Center for Security Policy, 7 December 2000.

Sakhuja, Vijay, Cdr, *Maritime Power of People's Republic of China: The Economic Dimension*, New Delhi, IDSA Fellows Seminar, 25 November 2000.

————, *The Military Component of Chinese Maritime Power,* New Delhi, IDSA Fellows Seminar, 28 July 2001.

From the Net

"The Chinese Missile Threat," available at http://www.cdiss.org/china_Essay.htm

Gill, Bates, "Chinese Military Power," Hearings on Chinese Military Power by the United States House of Representatives Armed Services Committee, 19 July 2000, available at http://www.brook.edu/views/testimony/gill/20000719.htm

————, "China's Weakness is What Makes It So Dangerous," *LA Times,* 6 April 2001, available at http://www.brook.edu./views/op-ed/gill/20010406.htm

Phadke, RV, *PLAAF- Shifting Airpower Balance: Challenges to India's Security,* CISAC Working Paper, February 2002 available at http://www.cisac.org/htm and http://www.stanford.edu

Record, Dr. Jeffrey, *"Thinking about China and War," Aerospace Power Journal,* (Winter 2001), Vol. XV, No. 4, internet version, at http://www.airpower.maxwell.af.mil.htm.

Singh, Swaran, "Sino-Indian CBMs: Problems and Prospects," *Strategic Analysis* article available at, http://www.idsa.org/an-jul-4.html

Transcripts of the *Record of Historic Richard Nixon- Zhou Enlai Talks in February 1972,* now declassified, Internet version at http://www.gwu.edu/~nsarchiv/nsa /publications/DOC _readers/kissinger/nixzhou/ index.html

Journals

Asian Defence Journal.

Asian Survey.

Aviation Week & Space Technology.

China Quarterly.

China Report.

The Economist.

Far Eastern Economic Review.

Foreign Affairs.

IDSA Strategic Analysis.

IISS: Military Balance.

India Today.

Indian Defence Review.

International Security.

Jane's Defence Weekly.

SIPRI Year Book.

Survival.

USI Journal.

The Week.

Newspapers

BBC FBIS Asia.

Chinese language newspaper reports and press releases.

Foreign Broadcasting Information Service, (FBIS) US Government.

The Hindu.

The Indian Express.

International Herald Tribune.

Los Angeles Times.

New Straits Times.

New York Times.

People's Daily.

The Pioneer.

PLA Daily.

South China Morning Post.

The Straits Times.

Stratfor online: Strategic Forecasting.

The Times of India.

Washington Post.

Xinhua-net and other Chinese newspapers online

Index

Notes

Notes

Notes

Notes